The Rotary Cement Kiln

Second Edition

by

Kurt E. Peray

Chemical Publishing Co., Inc.
New York, N.Y.

© **1986**

Chemical Publishing Co., Inc.

ISBN 0-82060-367-8

Printed in the United States of America

Preface

Often regarded as the heart of the plant, the kiln constitutes clearly the most important step in the process of cement manufacturing. It represents the largest single capital investment and consumes the major portion of the energy requirements in the plant. Regardless how much effort and attention is being given to the preparation of the kiln feed, the fact remains that the feed has to be properly burned in the kiln so that a good quality product can be sold to the customer. Because of its importance, the kiln burning operation deserves special attention and kiln operators should be properly trained. The old simple saying still holds true: "When the kiln discharges clinker, the company has a fighting chance to make some profits, but when no clinker is produced, no money can be made."

The rotary kiln requires specialized knowledge and experience on the part of the operator so he can successfully perform his job. Thus, with its complex instrumentation and multiple reactions, the kiln poses a significant challenge to the kiln operator. It is obvious that the kiln operator occupies one of the key positions in the production crew.

The Rotary Cement Kiln is the first handbook of its kind to deal not only with the theoretical aspect, but also with the actual control functions of kiln operation. First published in 1972, the original edition of this book dealt primarily with wet- and long dry-process kilns. Since that time, the cement industry has undergone a radical change brought about by the energy crisis of the mid-seventies. More fuel and labor efficient kilns were built to keep pace with the rapid advances in cement manufacturing technology. Capital that was sufficient twenty years ago to buy a complete new cement plant today will barely be enough to buy a kiln. But, the new modern preheater and precalciner kilns of today, outperform and outproduce the older wet and dry kilns by a wide margin. Most of these are also fully automatic, controlled by computers and the noisy, dusty burnerfloor of the past has been replaced by remote, air-conditioned control rooms. There is no question that these technological advances have benefited the kiln operator for they have made his job easier and more pleasurable. On the other hand, there is no doubt that the operator's responsibility has greatly increased because most are responsible not only for kiln operation but for the control of the raw and finished grinding departments too. It is the author's hope that this revised edition will be as well accepted in the cement industry as the first book. We have expanded each chapter to make this a more complete and up-to-date training and reference book not only for kiln operators but for supervisors and management staff as well. Most important, we have added extensive discussions for preheater and precalciner operations.

The author discusses the theoretical fundamentals, including basic cement chemistry, composition of the kiln feed, heat balances and heat transfer, combustion, flames, fuels, and the air circuitry in a rotary kiln.

Step-by-step descriptions of the control functions for the operation of a rotary kiln are extensively discussed. The described burning procedures and techniques have been tested over many years on kilns of various dimensions, and experience has proven them to be entirely successful. So much so that computer control programs have been recently written and successfully placed in operation that were based on the 27 basic kiln control conditions first introduced by the author in our first book. Adopted for hundreds of kilns worldwide, they are the foundation for stable and economical operations.

The appendix includes a section with conversion tables, definitions of common terms relating to rotary kilns, and a suggested outline for a training program for new operators.

Many thanks to Joseph J. Waddell who coauthored the first edition with me.

Contents

Part I

Kiln Systems and Theory

1.

History

Vertical furnaces and simple forms of shaft kilns were used for burning lime well over 2,000 years ago. History tells us that the Romans used a vertical furnace in which to burn a pozzolanic lime. Near Riverside, California are the remains of underground furnaces (Fig. 1.1) in which the early Mexican settlers burned limestone to make lime during the first part of the 19th century. In later times so-called bottle and shaft kilns were employed. Vertical kilns of the type shown in Fig. 1.2 were constructed in Southern California about the turn of the century.

Early development of the rotary kiln probably started about 1877 in England, but Frederick Ransome is usually credited with the first successful rotary kiln, which he patented in England in 1885. Although the first Ransome kilns were a major breakthrough in the cement industry at that time, many years passed before a successfully operating rotary kiln was put into production. It was mainly the pioneer work of American engineers a few years after Ransome's discovery that brought the concept of the rotary kiln out of its infancy. The first economical rotary kiln in America, developed by Hurry and Seaman of the Atlas Cement Company, went into production in 1895.

Shaft kilns with continuous feed are now used mainly and only for the burning of lime and minerals other than cement. Rotary kilns have replaced these shaft kilns entirely in the cement industry. Although years ago, shaft kilns showed lower thermal and power requirements than rotary kilns, the advent of the preheater and precalciner kilns with their increased output and fuel efficiency has apparently made the shaft kiln obsolete for the burning of cement clinker.

Fig. 1.1   Remains of underground furnaces that were used by early California Settlers for burning limestone to make lime. (*Riverside Division, American Cement Corp.*)

The first Ransome kilns were 45 cm (18 in.) in diameter and 4.5 m (15 ft) in length. Later, about 1900, the rotary kiln grew to 1.8 m (6 ft) in diameter by 18 m (60 ft) long which in todays terms would have to be classified as miniatures. Kiln sizes really started to explode in the 1960's when they reached dimensions up to 6.5 m (21 ft) diameter and up to 238 m (780 ft) length. With these enormous sizes and corresponding high output rates a considerable amount of new structural and control problems started to evolve. Refractory life in the kiln became uneconomically low, coolers couldn't handle that high output especially not during upset conditions, and mechanical equipment failures became weekly occurrences in many plants.

The energy crisis represented a blessing in disguise in matters of kiln design. Suddenly, fuel conservation became the number one priority item in most cement plants which led directly to increased construction of preheater kilns all over the North American continent. Although these pre-

Fig. 1.2   Vertical shaft kilns were commonly in use in the latter part of the 19th century. (*Riverside Division, American Cement Corp.*)

heater kilns satisfied the need for lower fuel consumption, they didn't meet the requirements for using low-grade fuel and ever-increasing demands for higher production rates.

In an attempt to gain these higher outputs, the Japanese cement industry increased preheater kiln sizes to a point where they were back to square one, namely, these kilns again became too large; frequent mechanical problems and short brick-life became the norm just as in the times of the dry and wet monster kilns. The major breakthrough came in Europe where precalcination was successfully attempted in the late 1960's using a very low bituminous shale as a component of the kiln feed in a conventional preheater kiln. Adding combustible materials to the kiln feed, at that time, was nothing revolutionary, for the author himself, in 1957, had burned a wet kiln in Canada that contained oil shale in the slurry. The European experience, however, was the first time such an addition was successfully tried in a preheater kiln and thus paved the way for today's precalciner kiln. Precalciner kilns are the latest advance in cement manufacturing technology. They combine low thermal requirements, are able to use low-grade fossil fuels or other combustible materials, and show output rates that were considered unattainable only a few years back.

2.

Types of Rotary Kilns

Generally speaking, the clinker manufacturing processes used in rotary kilns are classified into:

Wet-Process Kilns
Semidry Kilns
Dry Kilns
Preheater Kilns
Precalciner Kilns

Each of these types are discussed here.

2.1 WET PROCESS

Into this group fall all processes in which the kiln feed enters the kiln in the form of a slurry with a moisture content of 30 to 40%. In comparison with a dry-process kiln of the same diameter, a wet-process kiln needs an additional zone (dehydration zone) to drive off the water from the kiln feed. Therefore, it must be considerably longer in order to achieve the same production rate.

To produce an equivalent amount of clinker, a wet-process kiln requires theoretically more fuel than a dry-process kiln because of the extra heat required to evaporate the water. However, in actual operation of a kiln this fundamental fact does not always hold entirely true. As one progresses in the reading of this book, the reasons for these discrepancies between theory and actual operation will become clearer and understandable.

Advantages of a wet-process kiln are:

1. feed is blended more uniformly than in the dry process
2. dust losses are usually smaller, and
3. in moist climate regions, wet processing of the raw material is more suitable than dry because of moisture already present in the blend materials.

2.2 SEMIDRY PROCESS

This member in the group of rotary kilns is also widely known under the term *Grate Process Kiln* or *Lepol Kiln*. These kilns are as efficient in matters of fuel consumption as the most modern preheater and precalciner kilns. Output rates, however, lag behind the aforementioned types of kilns. However, it is advantageous to select a Grate Process Kiln over a preheater or precalciner kiln in places where raw material moisture is so high that it cannot be economically dried by waste heat from the kiln. Lepol Kilns, because of the fact that the kiln exit gases pass through the granular feed bed, operate with much lower dust contents in the waste gases which gives these kilns a decisive advantage over other preheater kilns. Instead of granulating the kiln feed, some plants use filter press cakes to feed the kiln. In such cases, the wet-kiln feed slurry is first passed through large presses for removal of the free water and more importantly, to remove alkalies before the filter cakes are fed to the kiln.

In the grate process, pulverized dry-kiln feed is first pelletized into small nodules by means of 10-15% water addition, then the nodules are fed onto a traveling grate where they are partly calcined before they enter the rotary kiln. Heating of the nodules is effected by the exit gases from the rotary kiln, the hot gases passing through the material bed from above as they are drawn downward through the grates by means of a fan. The partly calcined material then falls down a chute into the rotary kiln where final clinkerization takes place. Because the kiln feed is already partly calcined before it enters the kiln, the rotary kiln itself is only about one-third the usual length. Fig. 2.1 is a schematic diagram of the flow of gas and material through a Lepol grate-process preheater.

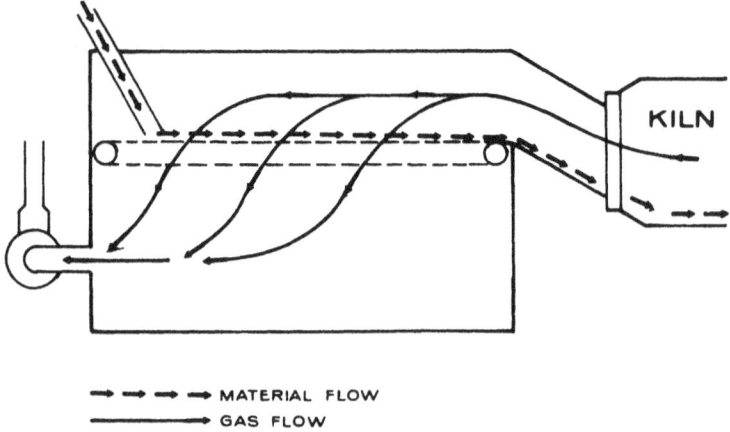

MATERIAL FLOW
GAS FLOW

Fig. 2.1   Flow diagram of a Lepol preheater.   Pelletized feed fed onto a traveling grate, is heated and partly calcined by hot kiln exit gases before it enters the kiln.

One advantage of grate-process kilns is the uniform size of clinker leaving the kiln, an aspect that is decidedly beneficial for grinding the clinker. However, there are some features not found in conventional rotary kilns that need very close attention; for example, production of the nodules and control of the thickness of the feed bed over the traveling grates. Such a kiln usually requires additional labor to attend the granulator plant.

2.3 DRY-PROCESS KILNS

As the term indicates, in this process the kiln feed enters the kiln in dry powder form. Dry-process kiln dimensions are similar to wet kilns in that they are long and typically show a length-to-diameter ratio of approximately 30:1 to 35:1. Dry-process kilns operate with a very high, back-end temperature and require watersprays at the feed end to cool the exit gases to safe levels before they enter the baghouse or precipitator. Most dry kilns are equipped with chain sections at the feed end to transfer heat, that otherwise would be lost, to the feed before the gases leave the kiln.

Fig. 2.2 shows a picture of a chain section. The gases enter the chains at a temperature of approximately 800 C (1470 F) and leave the kiln exit at a temperature of 450 C (840 F). In countercurrent flow, the material enters the chains with a temperature of 50 C (120 F) and emerges from the chain section with a temperature of 730 C (1350 F). Chain sections are a high maintenance item; difficult to repair but an absolute necessity for efficient operation. Because of the high costs of these chains, there is a tendency in many plants to neglect proper and frequent maintenance. However, it has been found in many instances that the costs saved, by not taking care of the chain system, are coming back manyfold in the form of higher fuel operating costs.

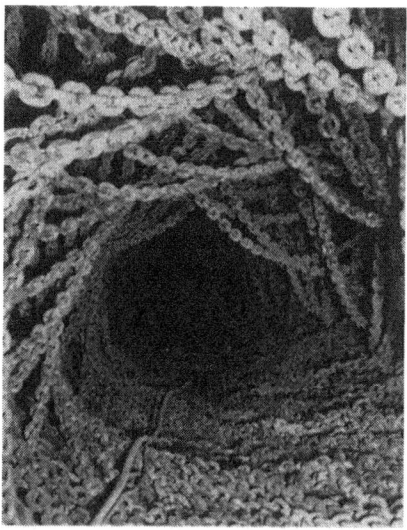

Fig. 2.2   The chain section of a kiln. The chains absorb heat from the hot gas and transmit the heat to the kiln feed.

There is an advantage found in dry kilns that none of the other kiln types exhibit. The high exit-gas temperature on these kilns renders them perfectly suitable for cogeneration of electrical power. As a matter of fact there are several plants with dry kilns that generate their own power and many existing plants are taking a hard look at the feasibility of adding a power plant to their facility. The reasoning is simply that generating power is energy-conserving and in some locations it may be more economical to add a power plant to an existing dry kiln than to convert this kiln to preheater status.

2.4   THE PREHEATER KILN

In the gas-suspension preheater kiln, the dry feed is preheated and partly calcined by the hot kiln exit gases in a tower of heat exchange cyclones. This concept is, contrary to popular beliefs, not new because a patent on this type of kiln was issued in Czechoslovakia already in the early 1930's. However, the suspension preheater kiln as it is known today did not come into its own right until after World War II when German kiln manufacturers were able to overcome the operational and structural problems of these types of kilns (Fig. 2.3).

The preheating of the kiln feed is done outside the rotary kiln proper, i.e., before the feed enters the kiln. The heat exchange between the gas and the material takes place in the cyclones while both are in suspension. Many different designs of preheater towers are in existence using this basic principle. The most common design is the parallel four-stage preheater. Some of these can reach output rates of up to 8000 metric tons per day. Exit gas temperatures leaving the top # 4 stage are around 340 C (640 F) and in many such plants, these waste gases are used (together in some cases with the waste gases from the clinker cooler) for drying and preheating of kiln feed in the raw grinding department. One drawback of preheater kilns is the high concentration of volatile constituents such as alkalies, sulfur, and chlorides in the kiln exit gases that give rise to numerous plug-up problems at the lower cyclone stage and kiln inlet. For this reason, most suspension preheater kilns must be equipped with an alkali and sulfur bypass that allows evacuation of a percentage of the kiln exit gases and thus bypasses the preheater cyclones. Such bypasses are not only used for reducing plug-up problems but in many plants are a necessity to keep the alkali content in the clinker below maximum permissible levels.

Suspension preheater kilns are the most energy-efficient types of kilns available operating typically with a specific fuel consumption of around 3138 MJ/ton clinker (750 kcal/kg, 2.7 MBtu/sh.ton).

2.5   THE PRECALCINER KILN

Roughly 15 years ago, Japanese cement manufacturers were confronted with the question of how best to increase the production rates of existing preheater kilns. As mentioned earlier, their preheater kilns reached limits

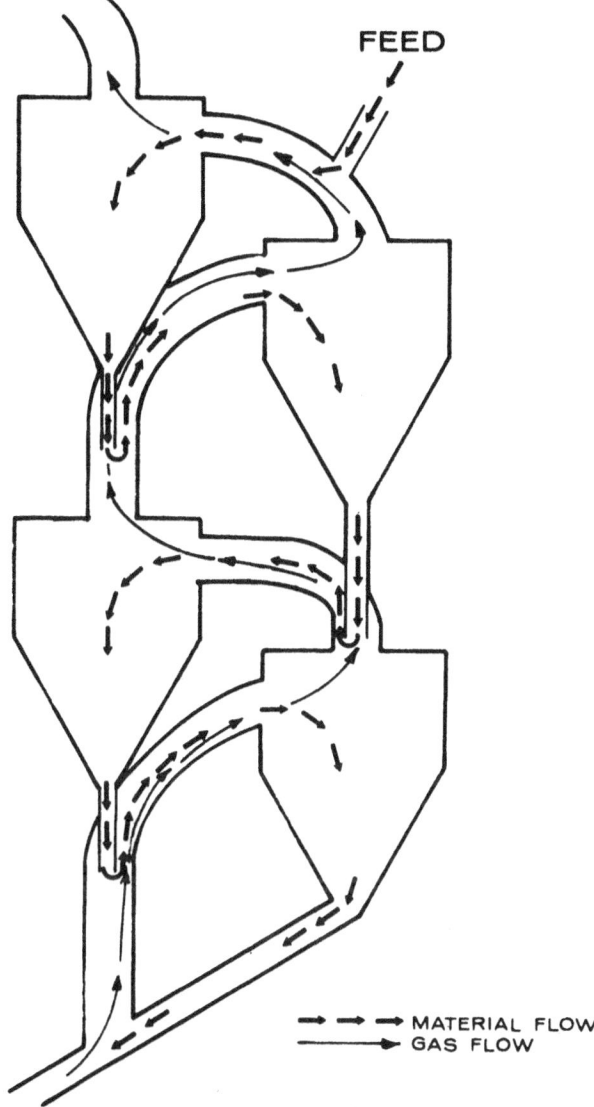

FEED

KILN

MATERIAL FLOW
GAS FLOW

Fig. 2.3 Flow diagram of a Humboldt gas suspension preheater, a multistage counterflow process which draws the gases through a series of cyclone collectors. The pulverized material, which is fed counter to the gas flow, becomes suspended in the gas stream and is heated in successive stages until discharged through the kiln-feed pipe.

in dimensions that gave rise to considerable operational and structural design problems. These kiln diameters became so large that refractory service life problems caused frequent and costly kiln down times. Then there was also the question of the possibility of using low-grade fuels in cement kilns that had to be addressed. Out of these considerations evolved the concept of auxiliary firing to precalcine the feed outside the rotary kiln. Precalciners are essentially suspension preheater kilns that are equipped with a secondary firing system (commonly referred to as flash furnaces) attached to the lower stage of the preheater tower. This allowed kiln manufacturers to construct smaller diameter kilns without sacrificing kiln output, for a precalciner produces 50-70% more clinker than a conventional preheater kiln of equal diameter. A precalciner, however, does not operate at a lower specific total fuel consumption than a preheater kiln; their consumptions are approximately the same. But by burning 30-50% of the total energy input at the rear of the kiln, the heat load in the burning zone proper is reduced producing a beneficial effect on refractory service life. Aside from this, being able to use low-grade cheaper fuels in the auxiliary firing unit results in a reduction of the costs per unit weight of clinker. Therefore, although the energy consumption is the same, the cost for this energy is usually lower—a decisive factor in favor of the precalciner kiln.

There are two different types of precalciner kilns—kilns with tertiary air ducts and kilns without. As is discussed later in more detail, combustion of any type of fuel requires a given amount of air, hence this air has to be provided to the auxiliary firing unit at the back end of the kiln. This is done in two ways, either, the air comes from the kiln itself (no tertiary duct) or it is supplied by the enormous excess waste gases from the clinker cooler by means of the tertiary air duct that runs parallel to the kiln. Precalciners without tertiary air ducts can be equipped with any kind of a cooler including planetary coolers whereas kilns with tertiary ducts cannot have planetary coolers. Kilns with tertiary ducts are more difficult to control for the operator since these kilns contain two distinct and separate combustion processes that must be closely controlled independently. However, the precalciner without the cooler air duct tends to be less efficient in fuel economy than the other when a large percentage of the kiln exit gases have to be bypassed in the preheater tower. Installation of a tertiary air duct is also very expensive and constitutes a high maintenance item.

In the short time span of 90 years, rotary kilns have undergone vast changes. Great improvements in fuel efficiency and control technology have been made and most newly constructed kilns are fully automatic and controlled by computers. With a rating of 55%, the modern efficient

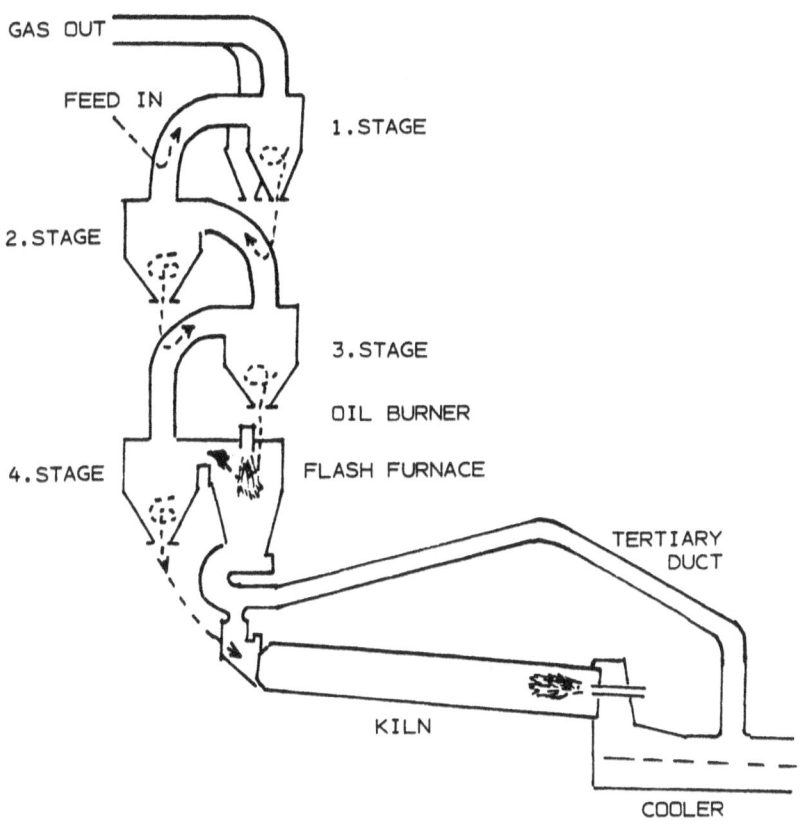

GAS OUT

FEED IN

1.STAGE

2.STAGE

3.STAGE

OIL BURNER

4.STAGE

FLASH FURNACE

TERTIARY
DUCT

KILN

COOLER

Fig. 2.4   SF—Suspension flash preheater kiln.

cement kilns appear to have approached the limit of best attainable fuel efficiency and output. Have kiln manufacturers reached optimum kiln design? Is the precalciner kiln the kiln of the future? The author and many cement engineers do not think so. Technology has never stood still in an age of rapidly accelerating industrial changes. Cement kilns will not be exempted from this trend. Although the precalciner kiln is an important piece of process equipment, it has numerous shortcomings such as the enormity of the tall tower on the kiln backend, and problematic

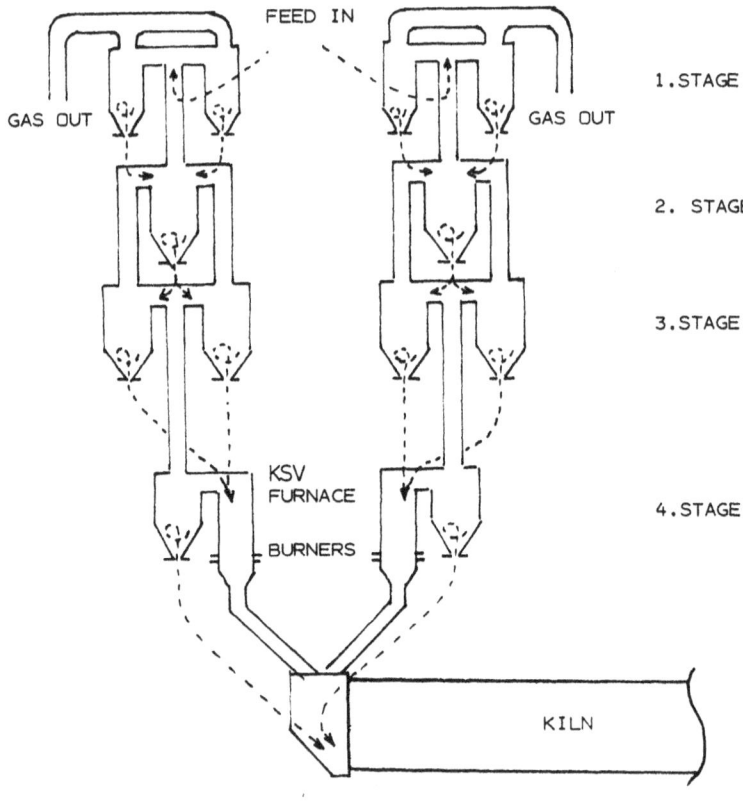

Fig. 2.5 KSV—Spouted bed and vortex chamber precalciner kiln.

environmental conditions. Because these systems have become so
sophisticated in matters of control and maintenance, well-trained specialists
are now an absolute requirement. There are many new concepts for
burning cement clinker that have been advanced by a multitude of forward-
thinking engineers. The fluid bed reactor, the separation of limestone
calcining from other raw material preheating, the two-stage traveling-grate
preheater are just some of the more "exotic" ideas that have come to the
forefront. Perhaps one of these will be the kiln of the future, so it is
important to keep an open mind toward such ideas and not regard them as
"not possible."

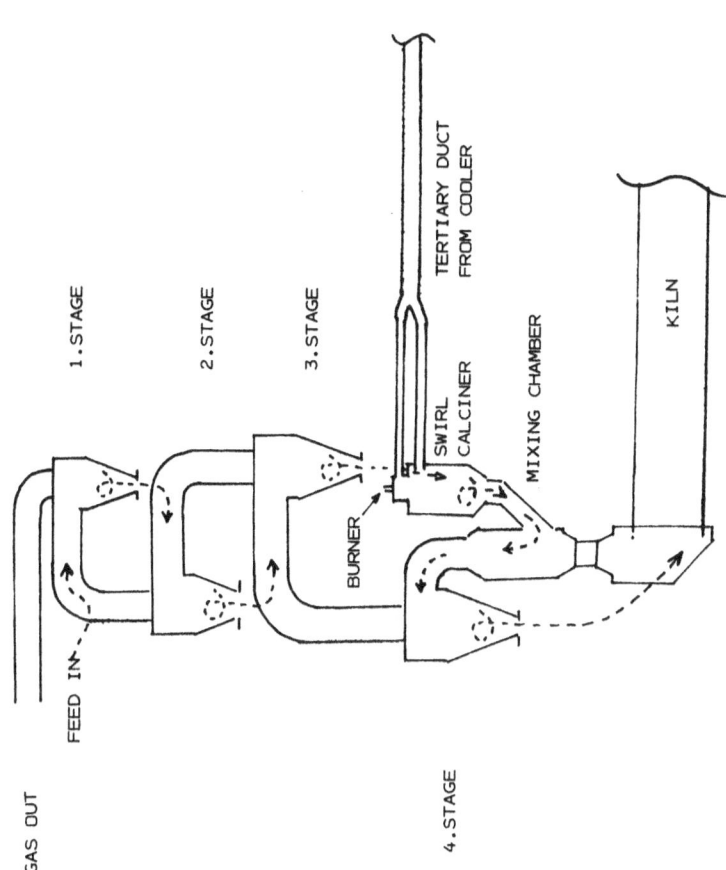

GAS OUT

1.STAGE

2.STAGE

3.STAGE

FEED IN

BURNER

4.STAGE

TERTIARY DUCT
FROM COOLER

SWIRL
CALCINER

MIXING CHAMBER

KILN

Fig. 2.6 RSP—Reinforced suspension preheater kiln.

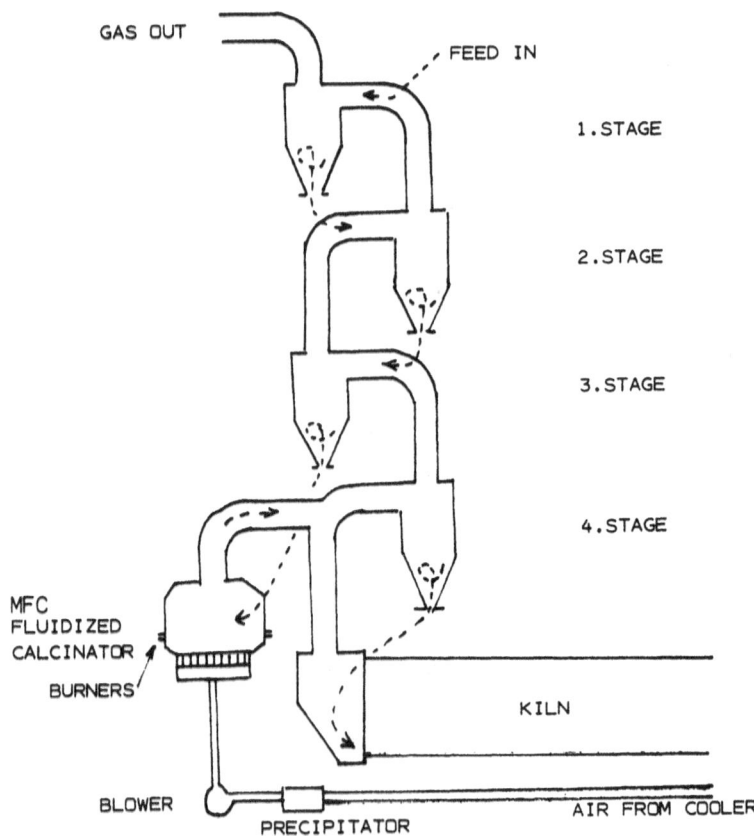

Fig. 2.7   MFC—Fluidized calcinator kiln.

3.

The Refractory

Because of the high temperatures existing inside a kiln during the clinker manufacturing process, it is necessary to protect the steel shell of the kiln with a refractory lining. If this protection were not provided, the shell would disintegrate within a few hours. A refractory is a material, usually nonmetallic, that is used to withstand high temperature. In a kiln, the refractory usually consists of brick of special composition and sizes as shown in Fig. 3.1. Some usage in recent years has been made of a cast lining continuously placed in a manner similar to placing concrete in a structure. In this method, the interior is progressively formed by means of special planks, welded anchors, and snapties. The kiln is rotated as necessary during placement of each section of lining so the workmen are always working at the same level.

Among kiln operators, refractory failure is considered the most critical upset in a kiln operation. Refractory failure inside the rotary kiln is indicated when the kiln shell becomes red hot because the refractory lining has either been entirely lost or has become so thin in an area that the kiln shell becomes overheated. Such a condition is dangerous because once the protection supplied by the refractory has been removed, the steel shell could easily be warped to such an extent that replacement of an entire kiln shell section becomes necessary. In most instances, however, damage can be avoided if the kiln is shut down for lining replacement as soon as the shell starts to show a large red spot. Because of the importance of such a situation, remedial procedures for hot shell conditions are described in Chapter 25.

Replacement of the kiln lining, especially in the burning zone, is unfortunately a frequent necessity, exerting a large strain on the operating

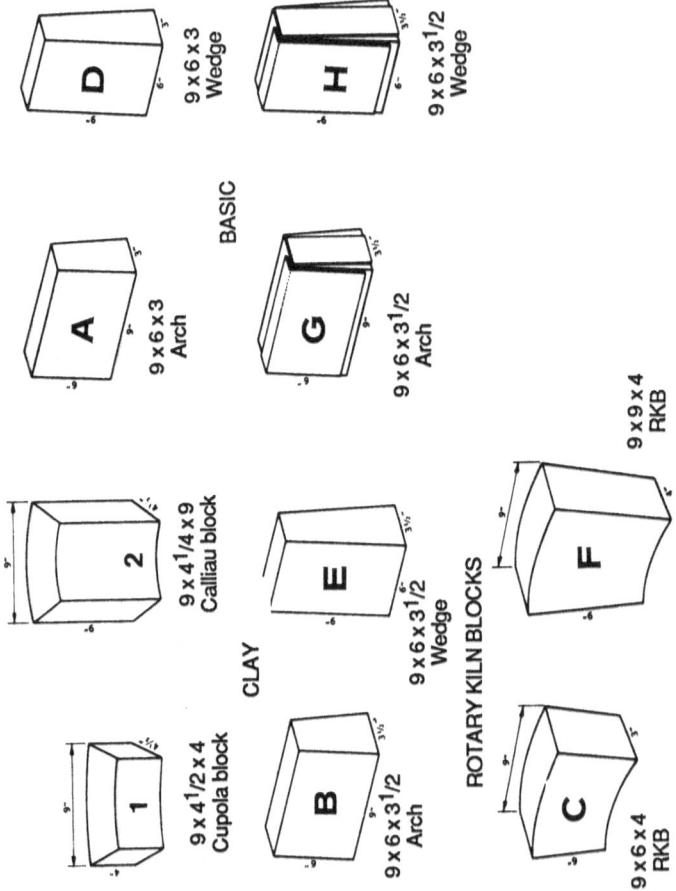

Fig. 3.1 Bricks for the refractory lining come in several sizes and shapes. (*Kaiser Refractories*).

budget and on production schedules. For example, replacement of an entire burning zone lining over a length of 50 ft in a 16-ft diameter kiln costs in excess of $80,000. This figure represents only the cost of the refractory itself and does not include the added expense of installation, nor the hidden cost resulting from loss of production and the extra fuel required to bring the kiln back to operating temperatures.

Ideally, one would like to obtain a service life on the burning-zone lining that would extend through a continuous operation cycle from one turn-around to the next. In other words, the optimum service life would be a period of either 11 or 23 months. In most kilns, however, this is more the exception than the rule. Although linings in the upper, cooler regions of the kiln can show lives of 5-20 years, in the intensely hot burning zone the life ranges from as little as 30 days to as much as two years. Regardless of the fact that all cement kilns operate within a narrow temperature range in the burning zone there still exists that large discrepancy between kilns in matters of refractory service life. The reason for this can be found in the fact that each kiln has its own specific characteristics and idiosyncrasies which greatly influence service life. The life of refractories in a kiln is governed by key factors which are discussed in the following section.

3.1 FACTORS AFFECTING WEAR OF REFRACTORIES

Frequency of Kiln Shutdowns.

Many plants have found that refractory life is often directly proportional to the number of kiln shutdowns that were experienced while the refractory was in the kiln. The more shutdowns and kiln stops, the shorter the life. The danger of damaging the refractory is directly related to the rate of cool-down of the kiln, the danger being the greatest when cooling is too rapid. The first step in preventing this situation is obviously to eliminate shutdowns by operating the kiln more efficiently on a continuous basis.

The second step is to make sure that cooling is slow and uniform when the kiln is shut down. Cooling time should be at least 8 h and preferably longer. A large guillotine damper to seal the kiln exit (back-end) helps to conserve heat inside the kiln and retards cooling during a shutdown. On some rotary kilns, retaining a small fire during the shutdown accomplishes more or less the same results. Kilns equipped with internal heat exchangers such as chains and crosses, however, are exempt from this procedure, because leaving a fire in such a kiln under these conditions might lead to chain failure due to the presence of oxygen at a fairly high temperature without any cold feed entering the kiln. To secure a uniform cooling, the kiln has to be turned and jacked on a regular schedule, because the feed bed and the refractory underneath it will cool much slower than the refractory which is exposed to the kiln gases. A suggested schedule for turning the kiln after a shutdown on both dry and wet process kilns of various lengths and diameters, given in Table 3.1, should ensure a uniform cooling of the refractory and kiln shell (the kiln shell contracts also during cooling).

TABLE 3.1

TYPICAL KILN JACKING SCHEDULE

Frequency of turns should be as follows:

1. Continuous rotation on auxiliary drive for 30 min.
2. One-third turn every 10 min for 1 h.
3. One-third turn every 15 min for 1 h.
4. One-half turn every 30 min for 4 h.
5. One-half turn every hour for 4 h.
6. One-half turn every 2 h for 12 h.
7. One-half turn once every 24 h regardless of length of shutdown.

NOTES: Each time before the kiln is turned, ascertain that no one is inside the kiln and first cooler stage. Once a shift, check the clinker loading in the cooler inlet and run this material out when the pile is higher than 2 ft. This is necessary because each time the kiln is turned a small amount of material is dumped into the cooler. During periods of heavy rainfall it may be necessary to jack the kiln more frequently to provide even cooling. This is a typical schedule for one particular kiln. Actual plant conditions determine the schedule for any other kiln.

Another procedure to secure slow cooling of the kiln is to make sure the draft fan (I.D. fan) is shut down immediately when the fire has been taken out of the kiln (an absolute must on a kiln with internal heat exchangers). The primary air fan should be left running only for such a time as needed to protect the burner pipe from heat during the early period of kiln cooling.

Too Rapid Heating of the Refractory.

Just as in rapid cooling, so heating the refractory too fast, can cause
thermal deterioration of the brick. The question of how much time should
be allowed to raise the kiln to normal operating temperature is subject to
wide differences of opinion. Recommendations by refractory manufacturers
vary to some extent depending on the type of refractory installed. One
factor sometimes overlooked is the kiln shell itself, which expands upon
heating. Governed partly by the thermal conductivity of the refractory,
this shell expansion takes place somewhat slower than expansion of the
bricks. Because of this, a period of at least 16 h should elapse when heating a kiln to operating temperature. This is on the conservative side.
After all, there is far less danger of damage when the temperature is raised
too slowly than when it is raised too fast. More time should be taken to
bring the kiln on line with thicker linings and larger kilns.

Overheating the Kiln.

One of the prime duties of the kiln operator is to make sure the kiln
never becomes dangerously overheated. When a kiln is overheated, the feed
starts to ball up and the coating turns to liquid in the burning zone. In a
matter of minutes, unless this condition is not drastically counteracted, the
entire refractory lining could be melted and lost as soon as the protective
coating has melted away. Overheated conditions are more apt to result
from erratic feed loading and advancement in the kiln rather than from any
action of the kiln operator. Indirectly, cycling and upset kiln operating
conditions can therefore be a cause for short refractory life. This then is
another reason why kiln operating stability is so important.

Quality and Uniformity in Size and Shape of the Refractory.

Since the costs of the refractories are low when compared to the costs
for the loss in kiln production, labor expenses, and fuel needed to reheat
the kiln, it is advisable to consider refractory quality and performance ahead
of the price of the refractory itself. On large diameter kilns it becomes
especially important that the refractory shapes do not deviate by more than
2 mm (0.078 in.) on each of its planes. Kiln managers also have to select
the right type of refractories for each location of the kiln which is easier
said than done. Experienced refractory salesmen that have a working
knowledge of rotary kilns can be of great help in this selection process.

Chemical Composition and Uniformity of the Kiln Feed.

A basic lining in the burning zone must have a good protective coating in order to achieve optimal service life. The type of coating is dependent largely on the chemical composition of the feed and its uniformity. Volatile constituents such as alkalies, sulfur, and chlorides can attack and weaken a refractory lining. In short, a plant chemist must not only concern himself with the ultimate cement quality a kiln feed will deliver but must also design the mix to possess good burnability and coatability properties in the kiln. This subject is discussed in more detail in the section on kiln-feed chemistry.

Mechanical Condition of the Tire and Kiln Shell.

Each kiln, with every revolution, undergoes shell deformations in the vicinity of the tire that have a detrimental effect on the lining. This deformation is generally referred to as shell ovality. There is no kiln that is perfectly round. Ovality can be checked with the "HOLDERBANK" shell test unit. Each plant should have a schedule to frequently check the slippage of the tire at least twice a month as there is a direct relationship between this slip and the shell ovality. Excessive slippage is an early warning signal that shell ovality might be too high. Of equal importance are annual checks of kiln alignment preferably when the kiln is in operation because misaligned kilns too, can result in excessive stresses on the lining.

Poor Location and Directed Flame Patterns.

Effect of these are extensively discussed in the chapter on flame control.

Installation Methods of Refractories.

Quality of workmanship in refractory installations has a profound influence on how well these refractories perform in a rotary kiln. A refractory liner can not achieve optimum life potential unless it has been properly installed in accordance with time-proven methods and procedures. Extra time and efforts exerted during construction of a lining can pay dividends in longer service life and kiln uptime.

Proper Kiln Operating Procedures.

This includes the whole gambit of operating procedures discussed in this book from combustion to burning zone control to kiln start and stop procedures. Again, applying time-proven procedures and being consistent in control follow-up can pay dividends in longer lining life. There are no short-cuts in this respect.

3.2 REFRACTORY REQUIREMENTS AND PROPERTIES

Certain requirements have to be imposed upon the refractory, depending on the conditions it will be exposed to in the area of the kiln where it is to be used.

Resistance to High Temperatures.

The refractory has to withstand the temperatures that can prevail under adverse conditions as well as those that prevail under normal conditions in the zone where it is being used. Not only does it require the ability to withstand high temperatures without melting, it also must maintain its structural strength at temperatures below its melting point, and must maintain a constant volume when exposed for prolonged times to the high temperatures.

Spalling Resistance

Any kiln shutdown, start-up, or severe operating upset usually creates large temperature changes in the kiln, and the refractory must possess the necessary shock resistance to withstand such temperature variations. Failure to possess this quality can cause the brick to crack. These cracks, generally referred to as spalling, develop in a horizontal direction on the brick.

Spalling results from thermal shock. The same reaction can be observed with a drinking glass: If we set a cold glass into very hot water the thermal shock causes the glass to crack. However, if we slowly raise the temperature of the water containing the glass, there is no thermal shock and the glass is not damaged. The same applies to the refractory in the kiln. When a cold kiln is fired, the temperature of the refractory must be raised very slowly to avoid spalling.

Resistance to Chemical Attack (Slag Resistance).

During the process of clinkerizing, ash, slag, and vapors formed during the combustion process can attack the refractory, reacting chemically with brick, depending on the type of fuel used. Furthermore, dust and alkalies entrained in the kiln gases can adhere to the bricks in the burning zone and react with the refractory. The ability of a refractory to withstand these chemical attacks is called its slag resistance. A brick that does not possess this resistance could be considerably weakened by chemical attack, resulting in premature refractory failure.

Abrasion Resistance.

Conditions encountered in a rotary kiln make it necessary that the refractory withstand the abrasive action resulting from the sliding kiln feed bed and also by dust entrained in the moving gas stream. This abrasion resistance is a prerequisite for all bricks installed in front of and behind the burning zone where coating is not usually formed.

Coatability.

One of the most important qualities required from the refractory in the burning zone where the highest temperatures exist, is its ability to take on a good coating and to hold this coating for a prolonged length of time. The importance of coating in the burning zone is discussed in Chapter 6. Just as the refractory acts as a protection for the kiln-shell, so the coating in turn acts as a protection for the refractory, thus serving to prolong the life of the brick in the burning zone.

3.3   TESTS OF REFRACTORIES

Selection of the best refractory for a given area in the kiln is not an easy task for the person who has to make this choice. This is especially true for a kiln in a new cement plant where no previous experience can be used as basis for this choice. To arrive at the proper selection, one has to deal with a multitude of refractory manufacturers and literally hundreds of different refractory types and shapes. Any one type of refractory can perform extremely well in one kiln and be almost useless in another. Because all kilns are different and many different raw materials are used to produce clinker, it is essential that the refractory is designed specifically for a particular kiln and its specific operating conditions.

Manufacturers of refractories usually furnish conventional information on their materials, such as compressive and tensile strength, modulus of elasticity, chemical analysis, thermal conductivity, density, porosity, and gas permeability. In addition, there are various special properties, determined by certain tests that have become standardized in the refractory industry. Results obtained from these tests, while not 100% conclusive, do furnish a good indication of the properties of the refractory and its resistance to various exposures within the kiln, and are the basis for the selection of a refractory particularly suited to any given area of the kiln.

Melting Point.

The melting point is the temperature at which the refractory starts to sag and lose its structure or shape. However, the given melting point by no means guarantees that the refractory will not start to sag at a lower temperature. Depending on whether a reducing or an oxidization atmosphere exists in the area where the refractory is used, and also depending on the composition of the refractory itself, melting can take place at a temperature below the theoretical melting point. Bricks in prolonged use and which have chemically reacted can have a melting point lower than indicated by the test.

For a fire clay refractory, a test is used (ASTM C24) in which the PCE (pyrometric cone equivalent) value is determined. PCE is the number of the standard pyrometric cone whose tip would touch the supporting plaque simultaneously with a cone of the refrctory material being investigated. Together with this value, the manufacturer usually states the corresponding melting point, which is the temperature at which the tested cone has softened and sagged to the same extent as a standard cone. Here again one has to remember that the maximum operating temperature should be well below the melting point for reasons previously stated.

Hot Load (Temperature of Deformation).

Certain refractories, when under load (when subjected to a given static pressure) undergo softening at a temperature far below the melting point. Softening of the bricks can take place because of their own weight or because of the pressure exerted upon them by the material bed in the rotary kiln. Overheating of refractory can also cause deformation and thus reduce the bearing strength under load. The test (ASTM C16) determines the resistance to deformation or shear of refractory brick when subjected to a specified compressive load at a certain temperature for a specified time. The results of the hot load are expressed as the temperature at which a definite deformation takes place when the brick is subjected to a given static pressure, or the percent deformation at a stated temperature when subjected to a given static load (usually 2 kg per cm^2 or 25 lb per $in.^2$). Because this test is performed over a relatively short period of time, it does not give an indication of the structural strength of the refractory when subjected to prolonged and constant high temperatures and loads. Such a prolonged exposure could weaken the refractory further.

Linear Expansion or Shrinkage (Reheat Test).

As stated above, a refractory, under prolonged service at high temperature and repeated heating, can undergo expansion or shrinkage. In the reheat test, these changes in volume are determined and generally referred to as irreversible (permanent) changes. The results are expressed in percent linear change at a stated temperature, a plus (+) sign indicating growth and a minus (—) sign, shrinkage. The test (ASTM C113) is conducted by placing the refractory specimens in a down-draft kiln for several hours, then cooling for at least 10 h, with length measurements made before and after test.

Panel Spalling.

Temperature changes during repeated heating and cooling can develop stresses in the refractory which in turn can cause cracking. In this test, the ability of a refractory to withstand repeated temperature changes is determined. The brick is heated to a given temperature on one side in a wall and then artificially cooled with a fan. Each brick is weighed before and after testing and the result is expressed in percent loss in weight. The basic procedure, a somewhat complicated test requiring special furnaces and control apparatus, is covered in ASTM C38. Briefly, it consists of heating a panel of bricks, at least 18 $in.^2$, then rapidly cooling the panel in a blast of air and water spray for a specified number of cycles.

Thermal Expansion.

In contrast to the reheat test in which permanent volume changes of a refractory are determined, the so-called reversible expansion of a refractory is indicated by this test. In other words the results, expressed in percent change in length, indicate to what extent a refractory expands when heated to any given temperature. Subsequent cooling of the refractory will cause it to return to its original dimensions.

ASTM vs. DIN.

All the discussions so far have been concentrated on testing methods in accordance with ASTM procedures, i.e., procedures as they apply to the North American refractory industry. But, foreign refractories are being used also in the United States, and their specifications and data are, in many instances, quite different. For this reason, a short discussion of the differences between the European and American methods for the more important tests is necessary.

European (DIN) testing is done on smaller cylindrical sample specimens (50 x 50 mm) whereas American testing is done on whole bricks. This clearly establishes an advantage for the American method because it allows observation of the behavior of an entire brick. But, at the same time the American method is more time consuming. The following list compares the two methods for specific tests.

Chemical Analysis: No difference between DIN and ASTM.

Apparent Porosity: Values given by DIN are usually slightly higher than ASTM results obtained.

Cold-Crushing Strength: No difference between DIN and ASTM results.

Pyrometric Cone Equivalent: No significant difference between the Seger (European) and the Orton (ASTM) test cones results.

Modulus of Rupture: DIN results tend to be higher than ASTM results.

Refractoriness Under Load (Hot Load Test): No correlation exists between these two methods because the testing methods are different from each other and, hence, no comparison can be made. DIN tests for both t_a, the starting temperature at which the sample is being deformed by more than 0.3 mm and for t_e, the temperature at which the sample is being deformed by more than 10 mm.

Reheat Test: Each method uses different temperatures at which the sample is being tested.

Panel Spalling: No comparison can be made between these test results because the methods are different from each other.

3.4 TYPES OF REFRACTORIES

Refractory replacements and problems occur mainly in the burning and the cooling zones of a rotary kiln, where the highest temperatures exist. Linings in the calcining and heating zones very seldom have to be replaced and usually do not represent a problem. For this reason, the following discussion is concerned with refractory for lining in the burning zone.

TABLE 3.2
RELATIONSHIP BETWEEN PROPERTIES OF A
REFRACTORY AND ITS RESISTANCE TO ATTACK

Property to be improved	ASTM test method	Required change in test result
Refractoriness	C24	Higher melting point.
		Higher PCE value.
		Decreased percentage of impurities.
		Decreased percent deformation.
Structural strength under load at high temperature	C16	Higher deformation temperature.
Volume stability at high temperature	C113	Decreased linear change.
Spalling resistance	C38	Decreased panel spalling.
		Higher thermal conductivity.
		Lower modulus of elasticity.
		Lower thermal expansion.
		Higher cold compressive strength.
Slag resistance		Higher density.
Abrasion resistance		Higher cold compressive strength.
		Higher density.

To indicate an improvement in the refractory property listed in Column 1, the test result should change as shown in Column 3.

The relationship between certain desirable characteristics of a refractory and properties indicated by tests is indicated in Table 3.2. This table can serve to assist in selecting a refractory. However, manufacuring methods, size and shape of the units, and the type of refractory all have an important share in influencing its ability to perform satisfactorily. For example, a high-alumina brick can be manufactured by either the dry press or the stiff mud method, and a brick made by the dry press process will have distinctly different properties from a brick of identical composition manufactured by the stiff mud process. Manufacturing methods affect the properties and behavior of basic refractories of equal compositions also.

For the purpose of this discussion, refractories commonly used in the burning zones can be classified into two main groups: the alumina-silica group and the basic group.

The Alumina-Silica Group.

As the title indicates, the two main components in these refractories are alumina and silica. With some limitations, an increase in the alumina content (which correspondingly reduces the silica content) will result in higher refractories.*

In addition to the improvement in refractoriness resulting from an increase in the alumina content, spalling resistance is improved (content of fluxing materials is decreased), conductivity and strength are enhanced and the refractory is more resistant to chemical attack. However, an important feature is that the high-alumina brick have a greater reversible expansion, a factor that requires special consideration during installation of the brick. Also, the slag resistance is lower than that of basic brick.

As stated previously, manufacturing processes affect refractory characteristics. Because of this a dry press high-alumina brick is preferred in applications where uniformity of size and good spalling resistance are important, but a stiff mud brick would be recommended when high strength and abrasion resistance are sought.

The Basic Group.

Basic refractories are manufactured mainly from periclase (a dense crystalline magnesia), dead burned magnesite, and chrome ore. For rotary kilns, the majority of basic bricks used fall into the magnesite-chrome

---

* Refractoriness: In refractories the capability of maintaining a desired degree of chemical and physical identity at high temperatures and in the environment and conditions of use (ASTM C71).

classification in which periclase makes up the largest portion of the composition, in contrast to chrome-magnesite brick with chrome ore as the dominant material. Basic refractories have a greater resistance to chemical attack from ashes, slag, etc. at high temperatures, but have a poorer spalling resistance compared to alumina bricks. Generally speaking, the periclase in a basic refractory is responsible for high refractoriness and volume stability, and chrome supplies spalling resistance and hot strength to the brick. Although not conclusive, the percentages of periclase and chrome give an early indication of spalling resistance and other properties of a basic brick. Basic refractories are preferred in the burning zone of a rotary cement kiln because they take on a coating more rapidly and hold the coating better than an alumina refractory. A coating properly formed over basic brick exhibits adherence characteristics quite different from those of a coating formed over a high-alumina refractory, as the coating will fuse with the surface layer of the basic brick but will only adhere without a strong bond to high-alumina brick. Because the coating is fused to the basic brick surface, there is a slight hazard to be considered, for if the coating is lost, part of the brick may be lost along with it.

The third type of refractory used in the burning zone is Dolomite bricks. These bricks are mainly composed of CaO and MgO and have a very close chemical affinity to the kiln feed. The big advantage of these liners is that they form a coating very rapidly once the kiln is brought to operating temperature. Because of their high heat-transfer coefficient, these bricks will show very high shell temperatures before such a coating is established and it is not uncommon to observe a faint dark red shell color in the early stages of a kiln start. The ideal location for placement of these types of liners is in the center of the burning zone and away from the kiln tires.

The CaO component in this brick, however, has a tendency to hydrate when exposed to moisture in the air. Care must be taken to make sure these bricks do not come into direct contact with moisture in storage and protective measures are necessary to shield the bricks from humidity during long kiln shutdowns. Some Dolomite brick manufacturers impregnate the bricks with tar to protect them during shipment and storage. Another manufacturer vacuum packs their product in airtight aluminum foil to achieve the same result. The thing to remember is that once a pallet has been broken open, the likelihood exists that these bricks will disintegrate in a short period of time.

When Dolomite linings are left in the kiln for a long period of time during a winter shutdown, proper precautions must be taken so that the lining remains unharmed. First, the coating should not be removed during this downtime. Second, the kiln feed present in the kiln at the time of the shutdown should also be left in and used to cover the coating/lining in the burning zone. This partly calcined feed acts as a dessicator to absorb any humidity present in the kiln. Third, the lining (coating) should be sprayed with diesel oil at weekly intervals to prevent penetration of moisture to the lining. The author knows of one cement plant that regularly shuts down the kiln for up to 5 months each winter and is successful in maintaining the integrity of the used, Dolomite lining. But, this appears more the exception than the rule. One might ask the question, why bother with these types of bricks at all? The advantages to their use are: the cheaper price of Dolomite bricks compared to their magnesite-chrome counterparts and the aforementioned rapid formation of new coating which is desirable in kilns that burn a tough, difficult-burning mix.

The newest type of burning zone liner for cement kilns is the Spinell-bonded brick. Recently introduced in the Japanese cement industry, this liner has shown some remarkable improvements in service life for it is reported to be as high as 1.5 to 2 times the life of high MgO-Cr liners. Spinell-bonded bricks are being offered now by almost every brick manufacturer and they all show chemical composition of around 10-15% alumina and 80-85% MgO. But these bricks are still in the development stages and each manufacturer's Spinell-bonded brick has its own characteristics as is the case with all other product lines. Prices for these bricks are correspondingly about 50–100% higher than conventional basic liners. Because of these high prices, Spinell-bonded bricks are normally installed in such places in the burning zone where everything else has failed.

3.5 REFRACTORY SHAPES

On the North American continent, rotary-kiln blocks, arches, and wedges are the most common refractory shapes used to line rotary kilns. In countries using the metric system of units, VDZ and ISO shapes are used. The following data will familiarize the reader with the dimensional differences between these shapes. It is important to note that dimension "a", i.e., the back cord, is the face of the refractory that is in contact with the kiln shell. All shapes are installed so that the given dimension "l" forms a parallel line to the kiln axis. Dimension "h" indicates the lining thickness.

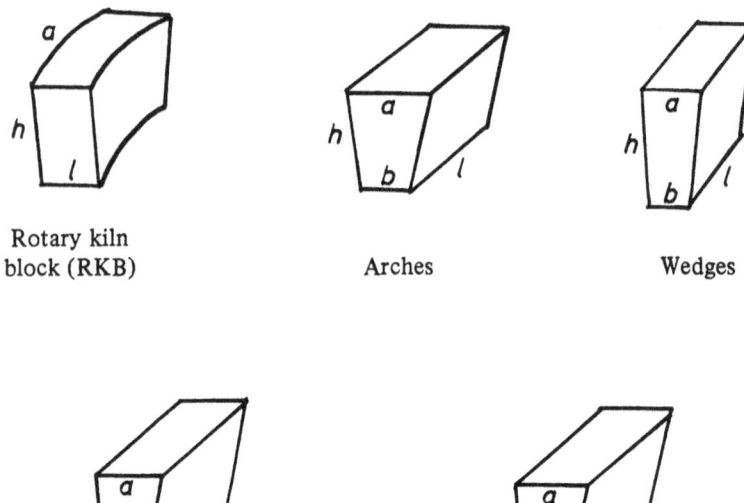

Rotary kiln
block (RKB)   Arches   Wedges

VDZ shapes   ISO shapes

TABLE 3.3
DIMENSIONS IN THE ENGLISH SYSTEM OF UNITS

(dimensions given in inches)

	RKB			Arches			Wedges			VDZ	ISO $\pi/3$
a	9	9	12	4	3.5	3	4	3.5	3	var.	4
b	v	v	v	v	v	v	v	v	v	var.	var.
c	v	v	v	v	v	v	v	v	v	2.8-4	
l	4	4	4	9	9	9	6	6	6	7.8	7.8
h	6	9	9	6	6	6	9	9	9	var.	var.

Note: VDZ and ISO shapes are usually manufactured for 6 3/10 in., 7 in., 7 7/8 in., 8 7/8 in., and 9 4/5 in. lining thickness (h).

TABLE 3.4
DIMENSIONS IN THE METRIC SYSTEM OF UNITS
(dimensions given in millimeters)

	RKB			*Arches*			*Wedges*			*VDZ*	*ISO π /3*
a	229	229	305	102	89	76	102	89	76	var.	103
b	ν	ν	ν	ν	ν	ν	ν	ν	ν	var.	ν
c	ν	ν	ν	ν	ν	ν	ν	ν	ν	72-102	
l	102	102	102	229	229	229	152	152	152	198	198
h	152	229	229	152	152	152	229	229	229	var.	var.

Note: VDZ and ISO shapes are usually manufactured for 160, 180, 200, 225, and 250 mm thick linings (*h*).

3.6 NUMBER OF BRICKS REQUIRED PER RING

Outside the United States, it is customary to use two different shapes of bricks with different backcords (a) to complete a full circle of the kiln circumference. Experience has shown that this produces a superior fit of the refractory to the shell particularly when the kiln shell is slightly out of round.

a) for RKB, arches and wedges

 for basic lining   *for aluminum brick*

$$n = \frac{12\ d\pi}{a + 0.059} \qquad\qquad n = \frac{12\ d\pi}{a + 0.039}$$

 n = number of bricks per ring
 a = back cord (in.)
 d = kiln diameter (ft)

b) for VDZ shapes

Consult manufacturer for number of bricks per ring of each shape required.

c) For I.S.O. shapes (π/3)

I.S.O. shapes have a uniform backcord of 103 mm. With an expansion insert of 1 mm, the cord length becomes 104 mm which is equivalent to π/3, explaining the reason for identifying these shapes by this nomenclature. With π a constant in the brick backcord and the circumference of the kiln shell, the calculation for the number of bricks required per circle becomes simple:

$$n = 1000D(0.0333) = 33.33D$$

where
  D = internal kiln shell diameter (m)

Example: How many bricks are required in I.S.O. shapes to complete a circle on a 4.8 m diameter kiln?
Answer: (33.33) (4.8) = 160 pieces.

3.7 NUMBER OF BRICKS REQUIRED PER UNIT KILN LENGTH

a) when dimension "l" is expressed in millimeters and kiln length in meters:

$N =$	Brick	RKB	Arches	Wedges	VDZ	ISO
	m	9.6n	4.3n	6.5n	5n	5n

b) when dimension "l" is expressed in inches and kiln length in feet:

$N =$	Brick	RKB	Arches	Wedges	VDZ	ISO
	ft	3n	1.33n	2n	1.524n	1.524n

n = number of bricks per ring

TABLE 3.5
KILN DIAMETER CONVERSION TABLE

ft	mm	ft	mm	ft	mm
8.84	2900	12.00	3937	15.54	5100
9.00	2953	12.19	4000	15.85	5200
9.14	3000	12.50	4101	16.00	5250
9.45	3100	12.80	4200	16.15	5300
9.50	3117	13.00	4265	16.46	5400
9.75	3200	13.11	4300	16.50	5414
10.00	3281	13.41	4400	16.75	5500
10.06	3300	13.50	4429	17.00	5577
10.36	3400	13.72	4500	17.07	5600
10.50	3445	14.00	4593	17.37	5700
10.66	3500	14.02	4600	17.50	5742
10.97	3600	14.32	4700	17.68	5800
11.00	3609	14.50	4757	18.00	5900
11.25	3691	14.63	4800	18.29	6000
11.27	3700	14.93	4900	18.50	6069
11.50	3773	15.00	4922	18.59	6100
11.58	3800	15.24	5000	18.90	6200
11.89	3900	15.50	5085	19.00	6234

3.8 REFRACTORY CONSUMPTION REPORTING

Finally, mention must be made of methods of reporting refractory consumption. It is easy to fall into the trap of reporting refractory service life in terms of days, months, or years. Unless the kiln has operated continuously during this period of so-called "days service life" there is not much use in comparing such figures with other periods or kilns. A kiln manager should, if possible, find the time to express refractory consumption in terms of kilograms refractory per ton of clinker (or lb/sh.t.Cl.) produced. More paperwork is required but the data thus obtained is much more meaningful to the study of individual refractory types. Likewise, proper record keeping on brick charts is an absolute requirement for any plant that is seriously concerned about improving refractory service life.

References

1. Z. Clausen, C.F. Jan., 1962 "The Evolution of the Cement Kiln, a Historical Sketch," *Journal, PCA Research and Development Laboratories.*
2. Budnikov, P.P. "The Technology of Ceramics and Refractories" *M.I.T. Press,* Mass. Inst. of Technology.

4.

Fuels

Fuels used in burning a rotary kiln are classified into three groups: solid, liquid, and gaseous fuels. In many kilns a combination of two different fuels are used.

4.1 SOLID FUELS

Solid fuels used in rotary kilns are generally of the following types and properties:

Group	Moisture (%)	Volatile matter (%)	Fixed carbon (%)	Ash (%)	Heating value	
					kj/kg	Btu/lb
Anthracite	2.3	3.1	87.7	6.9	31,508	13,540
Semianthracite	2.2	12.4	67.4	18.0	28,435	12,270
Bituminous low-vol	3.5	18.2	74.4	3.9	33,858	14,550
Bituminous med-vol	3.1	25.0	66.8	5.1	33,253	14,290
Bituminous high-vol A	2.6	30.0	58.36	9.1	31,670	13,610
Bituminous high-vol B	8.2	36.1	48.7	7.0	28,296	12,160
Bituminous high-vol C	12.1	40.2	39.1	8.6	26,714	11,480
Subbituminous A	16.5	34.2	38.1	11.2	22,665	9,740
Subbituminous B	23.2	33.3	39.7	3.8	21,920	9,420
Subbituminous C	24.6	27.7	39.9	7.8	20,035	8,610
Lignite	34.8	28.2	30.8	6.2	16,778	7,210
Delayed coke		13.0		1.2	33,974	14,600
Fluid coke		5.0		2.0	34,440	14,800

Subbituminous coal, lignite, oil shale, and other low-grade, solid fuels are fired predominantly in the flash calciner since these low heating-value fuels are unsuitable to sustain good combustion conditions in the burning zone. As a matter of fact, one of the most important advantages of a pre-calciner kiln is that low-grade fuels can be used for firing in the auxiliary furnace of the precalciner. Petroleum coke is used either at the precalciner furnace or intermixed with other fuels in the burning zone. Several pre-heater kilns in various parts of the world are being used to dispose of old automobile tires, others use wood chips, and even garbage for introduction to the back end of the kiln. This method serves two purposes simultaneously. It does away with an environmental problem of waste material disposal and at the same time delivers valuable heat to the kiln which otherwise would have to be suppplied by the more expensive conventional fuels. Petroleum coke possesses a much higher ignition temperature [≈590 C (1100 F)] and is therefore suited for intermixing with raw feed in a suspension preheater kiln. In effect, this method upgrades a preheater to precalciner status without the need for installation of an auxiliary firing system at the preheater tower. There is one drawback in using petroleum coke in any rotary kiln, namely its unusually high sulfur content.

One of the inherent disadvantages of coal is not only its explosive potential when finely ground but also its tendency to ignite spontaneously while in storage. High-volatile, high-sulfur, high-moisture coals are especially vulnerable to such ignition. Special precautions have to be taken when stockpiling coal to prevent undesirable temperature rises in the pile. Common precautions taken as standard procedures are:

a) Building a coal pile in layers of 2–3 ft and compacting these layers with a bulldozer or roller before the next layer is applied.
b) No foreign materials such as rags, pieces of wood, or paper should be present and the storage area should contain a concrete floor with proper drainage.
c) Testing wells (closed-ended steel pipes inserted into the pile) should be strategically spaced over the entire pile to allow for temperature monitoring by means of thermocouples or thermometers.
d) For long-term, strategic, coal reserve piles, coal should be screened with the maximum amount of fines removed, i.e., store only coarse lump coal if possible.
e) When a smothering "hot spot" is detected, do not apply water in

an attempt to put the "fire" out. Instead, dig it out and repack the pile.

Coal firing on rotary kilns is done by either a direct-, semidirect-, or an indirect-fired system. In the direct-fire system, the coal is dried, ground, and immediately conveyed into the burning zone for firing (see Fig. 2.1). Grinding is being done in an air-swept bowl or roller mill with the particle size classification being done within the mill itself. Hot, excess cooler air, tempered down to safe levels is used for drying and conveying of the coal. This system is normally quite safe, easy to control but has the drawback that a large amount of relatively cold primary air must be used for firing. This makes this system less efficient than an indirect-fired one. It also limits the capability of the operator to change the flame shape at will. Most wet- and dry-process kilns, i.e., the older types, are being fired by this method.

Semidirect-fired systems mostly use an air-swept ball mill from where the pulverized coal is conveyed to a small cyclone (holding bin) above the primary air pipe. The coal then discharges through a variable-speed rotary feeder directly into the primary air pipe from where it is being blown into the kiln (see Fig. 2.2). Since these holding bins are usually relatively small, the mill, just as in the case of the direct-firing system, must be operating at all times while the kiln is running. When compared to the direct system, semidirect firing requires less primary air to be used, can operate at higher primary air temperatures, and can still supply the kiln with coal when the coal mill is shut down for up to 10 min.

In the indirect-fired system the coal is dried and ground in a roller or ball mill system that is separate from the actual kiln firing system. The ground coal is stored in a large silo from where it is withdrawn at metered rates to the primary air pipe (see Fig. 2.3). This grinding system does not have to be operated continuously as the capacity of the coal silo is usually sufficiently large to hold a supply of coal for up to 10 h. This is the type of system that most new cement kilns employ. Several kilns can be fired from this single grinding plant and coal rates, to the kiln, can be adjusted instantaneously by the operator. It also allows for optimum primary airflow control. This indirect system is the most complex of the three systems. It needs special controls, safety devices, and a dust (coal dust) collector because of the presence of vast quantities of finely ground coal.

It is important to strongly emphasize to kiln operators how important it is to keep close control over any coal-firing system. These systems have been designed for safety, are safe, and can remain so provided they receive the proper attention and maintenance. There are no shortcuts that can be used here: standard operating procedures must be fully understood and adhered to for the simple reason that fine pulverized coal can be an ingredient for a highly explosive mixture when certain conditions prevail. There is no reason to have any fear of these systems as long as the operator fully understands what must be controlled and knows exactly what to do when something goes wrong. An operator must show respect toward coal firing and should periodically review the operating procedures to have himself prepared for any eventuality. Most important of all, if an operator has some doubts about the accuracy of any instrument reading in the coal grinding and firing system, or when he knows something is not functioning properly, he must have it fixed right away.

4.2   LIQUID FUELS

Liquid fuels used for combustion in a rotary kiln are almost exclusively of the Bunker B or Bunker C class. These are residual oils from refineries after the more volatile products in the oil have been removed. Because both types fall into the class of heavy oils, they have to be preheated before they can be pumped and atomized. Steam plants usually deliver the necessary heat to the heat exchangers where the fuel is raised to the temperature that ensures the correct viscosity.

To obtain good combustion, fuel oil has to be atomized, which means the oil has to be broken into small droplets to promote easy combination with the oxygen. This is done by means of an atomizer nozzle in which the oil is forced through an orifice at high pressure, creating a turbulent motion. The orifice size determines the pressure of the fuel at the burner-tip, which in turn influences the flame shape. A small orifice results in higher fuel pressure and less fuel passing through the burner than a large orifice. Because certain high and low limits are set for a given orifice, it is often necessary to exchange the orifice for one with a different size of opening when a large change in fuel rate is required.

Fuel pressure is of utmost importance for the shape of the flame and consequently has to be frequently checked by the kiln operator. If the pressure, for example, is too low, complete combustion of the poorly atomized oil particles cannot take place, resulting in impingement of the unburned fuel on the coating and the feed bed.

4.3 GASEOUS FUELS

Although other gaseous fuels such as liquefied petroleum gas, hydrogen and coal gas are used in some industries, gaseous fuel firing of rotary kilns in the cement industry is done almost exclusively with natural gas, one reason being that it is the cheapest of all gaseous fuels. Natural gas offers several advantages for the kiln operator, as it needs no preparation such as drying, grinding, or preheating, and combustion takes place readily once the gas has been mixed with the proper amount of air and ignition temperature has been reached. The burning zone atmosphere is much "cleaner" because a natural gas flame is not as luminous as an oil or coal flame. A natural gas-fired burning zone appears "hotter" to the operator than one with other fuels, making a short adjustment period for the kiln operator necessary whenever firing is switched from another type of fuel to natural gas.

Another important advantage of natural gas firing is the need for very little or no primary air, thus more valuable hot secondary air can be used for combustion in the kiln.

4.4 COMPOSITION AND HEAT VALUE

A basic knowledge of heat generation and fuel efficiency is essential for any kiln operator, who should remember that the cost of fuel is one of the major expenditures in the manufacturing process of cement. Efficient use of the fuel depends in a large measure on the kiln operator's skill.

Composition and heat value of fuel are of significant importance to everyone concerned with operation of a rotary kiln, as they are the basis for heat transfer calculations and for investigation of the fuel efficiency of a kiln. Fuels differ considerably in the amount of heat they give out when burned, hence it is desirable to measure their heating values by means of an instrument called a calorimeter, expressing the heating value as the number of calories or British thermal units (Btu) given off when a measured amount of fuel is burned. One gram-calorie, or small calorie, is the amount of heat required to raise the temperature of one gram of water one degree centigrade. Similarly, a kilogram-calorie, or greater calorie, is the amount of heat required to raise the temperature of one kilogram of water one degree centigrade. In the English units, one Btu is the amount of heat required to raise the temperature of one pound of water one degree Fahrenheit. In North America, heating value of fuels is usually expressed as the number of Btu per pound for coal, Btu per pound or gallon for liquid fuels, and Btu per cubic foot for gaseous fuels. See Table 4.1. Metric units may be either gram-calories or kilogram-calories per kilogram or liter. Conversion factors are given in the appendix.

TABLE 4.1
COMPOSITION AND HEATING VALUE OF FUELS

SOLID FUEL
Pulverized coal

Carbon	Hydrogen	Oxygen	Nitrogen	Sulfur	Ash	Volatile matter	Btu/lb
78%	5%	6%	1.5%	1%	8.5%	30%	13,000

LIQUID FUEL
Bunker oil C

Carbon	Hydrogen	Oxygen	Nitrogen	lb/gal	Btu/gal	Btu/lb
86%	12%	1%	1%	8.1	146,500	18,700

GASEOUS FUEL
Natural gas

Methane	Ethane	Nitrogen	Other hydrocarbons	Btu/ft^3
80%	15%	1%	4%	1100

With the recent world-wide (except USA) standardization of the system of units, fuel heating values are now expressed in kiloJoules per kilogram (kJ/kg) or per cubic decimeter (kJ/dm³). Conversion factors are given in the appendix.

Divergent views still exist about the manner in which specific fuel consumptions are expressed. If it is stated that a particular kiln consumes 3722 kJ/kg of clinker (3.2 million Btu/sh.ton), then more information would be necessary. Determination of whether the gross or net heating value of the fuel was used and whether the fuel consumption was under steady-state kiln operating conditions or based on month-end fuel inventory figures would have to be made. If the cement industry would adopt the "NET HEATING VALUE/STEADY STATE" method, then efficient and accurate determinations of fuel consumption could be made. Gross heating values are the heat produced by a unit quantity of coal in a bomb calorimeter. The net heating values are based on both the heat produced at constant atmospheric pressure and the water in the combustion product remaining in vapor form. Net heating values can be calculated from the results of the calorimeter test by subtracting the hydrogen component as follows:

a) for Metric units

$$Hu_{net} \; = \; (Hu_{gross} - 5150\ H_2)\ 4.187 \qquad (kJ/kg)$$

b) for English units

$$NET\ value \; = \; GROSS\ value - 99270\ H_2 \qquad (Btu/lb)$$

The gross heating value is approximately 3.5% higher than the net heating value. It is quite possible that a particular kiln might show a specific fuel consumption of 4886 kJ/kg (4.2 MBtu/sh.t.) in steady-state condition but the month-end figure reported might show this as 5375 kJ/kg (4.62 MBtu/sh.t.) because these month-end figures include the clinker and fuel losses due to frequent kiln stops and starts or material losses that might not be associated at all with the operation of the kiln itself. It is easy to get on the wrong track in the matter of fuel conservation when coal and clinker storage areas are inadequate to prevent these materials from being blown by the wind over the countryside, or washed down the road during heavy downpours. These areas can not be overlooked in a fuel conservation program. It is therefore essential to check the specific fuel consumption under steady-state conditions and then compare this with the specific fuel consumption that is derived from the month-end inventory figure. If the latter is higher than 3% from the former, it is necessary to take a serious look at the existing clinker and fuel handling and storage systems. Attention to this can save the company many thousands of dollars every year.

5.

Combustion

Before entering into a detailed discussion on the combustion conditions in the kiln, it is appropriate to review some fundamental laws related to gases and combustion. Knowledge of these laws are essential in order for any kiln operator to understand the duties involved in his job. When malfunctions and unusual problems occur, such as chain fires, rapid and unusual temperature rises in the fuel or firing system, an understanding of these laws combined with common sense can make the difference between a potentially serious, out-of-control condition and a safe solution.

5.1 GAS LAWS

All gases, including the hot gases in a kiln, behave in a certain manner under external influences. A perfect gas is one that obeys very closely certain physical laws. For all practical purposes, the gases under consideration in this chapter may be assumed to be perfect gases. Adjustments to the air circuitry of a rotary kiln are part of the responsibility of the kiln operator, hence a basic knowledge of these laws is necessary to assist him in making the correct adjustments.

Before entering into a discussion of the gas laws, the reader is reminded that pressures and temperatures used in the following equations are *absolute* . Absolute values are determined:

a) by adding 460° to the temperature in degrees Fahrenheit, e.g., 60°F = 520° absolute.

b) by adding atmospheric pressure to gauge pressure; while atmospheric pressure varies with altitude and weather conditons, it is sufficiently accurate for these computations to use 14.7, i.e., 50 psi gauge = 64.7 psi absolute.

In a like manner, CGS units are obtained by adding 273 to the temperature in degrees centigrade, and 76 to the pressure in centimeters of mercury.

Boyle's Law.

In 1662 Robert Boyle expressed this law as follows:
Under constant temperature, the volume of a given mass of gas varies inversely as the pressure upon it. In other words, pressure times volume, at constant temperature, is a constant expressed mathematically

$$P_1V_1 = P_2V_2 \qquad (5-1)$$

where:
P_1 = initial pressure (absolute)
P_2 = final pressure (absolute)
V_1 = volume under P_1
V_2 = volume under P_2

Example: The volume of a certain gas, when measured by an instrument indicating a barometric pressure of 72 cm of mercury, is 43,000 ft^3. What will be the volume at standard presure (76 cm Hg) while the temperature remains constant?
Substituting in Eq. (5-1)

$$72 \times 43,000 = 76 \times V_2$$

$$V_2 = 43,000 \, \frac{72}{76} = 40,737 \text{ ft}^3$$

Charles' Law.

Similar to the discovery of Boyle, Jacques Charles in 1787 discovered that the volumes occupied by a given mass of gas at different temperatures are proportional to the absolute temperatures of the gas provided the pressure remains constant.

All gases expand when the temperature is raised. If the temperature of a gas is raised 1 F, it will expand 1/460 of its original volume. In CGS units, a temperature rise of 1 C will result in an expansion of 1/273 of the original volume.

Mathematically

$$\frac{V_1}{T_1} = \frac{V_2}{T_2} \qquad (5\text{-}2)$$

where:

T_1 = initial temperature, absolute
T_2 = final temperature, absolute
V_1 = volume under T_1

Example: The volume of a gas when measured at a temperature of 450 C is 5700 m³. What will its volume be at 520 C?

Reduce the Celsius reading to absolute (Kelvin) temperatures:

$$450\ C = 450 + 273 = 723\ K$$
$$520\ C = 520 + 273 = 793\ K$$

$$\frac{5700}{723} = \frac{V_2}{793}$$

$$V_2 = \left(\frac{5700}{723}\right) 793 = 6252\ m^3$$

Gay-Lussac's Law.

Finally, there was Joseph Gay-Lussac who, early in the 19th Century, delved further into the action of gases, and developed the law that bears his name, which states that the pressure exerted by a given mass of gas will increase in proportion to the temperature if the volume is held constant. All units again are in absolute values.

$$\frac{P_1}{T_1} = \frac{P_2}{T_2} \qquad (5\text{-}3)$$

in which pressure and temperature units are as previously described.

Example: The pressure of a gas when measured at 900 F is 72 cm of mercury. What will be the pressure if the temperature is raised to 1050 F and the volume remains constant?

Reduce the Fahrenheit readings to absolute temperatures:

$$900\ F = 900 + 460 = 1360\ A$$
$$1050\ F = 1050 + 460 + 1510\ A$$

Substituting the values in Eq. (5-3)

$$\frac{72}{1360} = \frac{P_2}{1510}$$

$$0.05294 \times 1510 = 79.9\ cm\ Hg$$

A General Law.

Now reviewing the above three basic gas laws, it becomes apparent that a general law can be stated, based on all three of the basic laws. The general equation is:

$$\frac{P_1 V_1}{T_1} = \frac{P_2 V_2}{T_2} \qquad (5\text{-}4)$$

in which P_1, V_1, and T_1 are the original pressure, volume, and temperature, and P_2, V_2, and T_2 are the final values, with temperatures and pressures expressed in absolute units.

Example: A gas measured 45,000 ft^3 at 900 F under a pressure of 75 cm Hg. What will be the volume at 1050 F at a pressure of 76 cm?

First reduce the Fahrenheit readings to absolute:

$$900\ F = 900 + 460 = 1360\ A$$
$$1050\ F = 1050 + 460 = 1510\ A$$

Substituting in Eq. (5-4)

$$\frac{75 \times 45,000}{1360} = \frac{76 \times V_2}{1510}$$

$$V_2 = \frac{75 \times 45{,}000 \times 1510}{76 \times 1360} = 49{,}290 \text{ ft}^3$$

These gas laws can help a new operator to understand some of the fundamentals in kiln control. They are the solutions to some common basic control problems an operator might be confronted with on an almost daily basis.

5.2   THE COMBUSTION REACTION

Successful operation of a rotary kiln requires an adequate source of heat that will first raise the kiln to the desired operating temperature, and will then maintain this temperature by compensating for the various heat losses occurring in the kiln system, including the heat required for the process. The required heat is obtained by combustion of the fuel, a chemical rection in which carbon, C, hydrogen, H_2, and sulfur, S, in the fuel combine with oxygen, O_2, in the air. To obtain combustion, two requirements must be fulfilled:

a) sufficient oxygen must be present to mix with the fuel, and
b) a certain temperature must be maintained to ignite the fuel-oxygen mixture.

An operator must always remember the triangular relationship that leads to combustion:

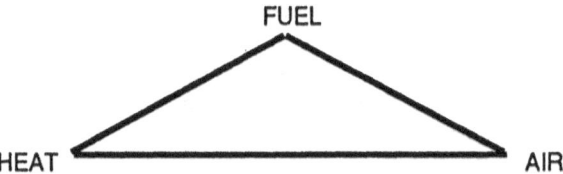

If any one of the three links is missing, no combustion will take place. If there is not enough air in the kiln, there will be no proper fire. Likewise in a dry or wet kiln, too much heat and air in the chain section will result in chain fire because chains contain carbon which acts as fuel. In order to stop a fire, the elimination of one of the components of the triangle is necessary, e.g., choking off the oxygen (air) supply. This will be discussed in greater detail later on. In addition, firing conditions inside the kiln must be such that the fuel particles undergo complete combustion while the fuel is still in suspension in the kiln atmosphere.

5.3   THE STANDARD COAL FACTOR, COMBUSTION AIR REQUIREMENTS

To determine the approximate combustion air needed to burn a given unit weight of coal, the formulas given below can be used when no ultimate coal analysis is available. The combustion air requirements here include 5 percent excess air.

a) English units   *b) Metric units*

$$SCF = \frac{100 - a}{100} \cdot \frac{b}{12,600} \qquad\qquad SCF = \frac{100 - a}{100} \cdot \frac{B}{7000} \qquad (5\text{-}5)$$

lb air/lb coal = 10.478 *SCF*   kg air/kg coal = 10.478 *SCF*

where:
   SCF = standard coal factor
   a = percent moisture in coal (as fired)
   b = heat value of coal (Btu/lb as fired)
   B = heat value of coal (kcal/kg as fired)

Fuel ignition temperatures, the third component in the combustion triangle, are approximately:

	°C	°F
for coal	250	480
for fuel oil	200	400
for natural gas	550	1050
for petroleum coke	620	1150

Here it can be seen why gas- and coke-fired kilns tend to have their ignition point of the flame much further back in the kiln than coal- or oil-fired kilns. In addition, combustion in a kiln requires that sufficient time must be available to accomplish complete combustion while the fuel is in suspension in the kiln atmosphere.

When complete combustion takes place, carbon dioxide (CO_2), water vapor (H_2O), and sulfur dioxide (SO_2) are formed:

$$C + O_2 \rightarrow CO_2$$
$$2H_2 + O_2 \rightarrow 2H_2O$$
$$S + O_2 \rightarrow SO_2$$

These equations for complete combustion indicate that when fuel is properly burned, 1 part of carbon will combine with 1 part of oxygen to form the combustion product carbon dioxide. However, when incomplete combustion takes place, then instead of carbon dioxide the combustion product is carbon monoxide:

$$2C + O_2 \rightarrow CO$$

Oxygen needed for combustion originates from the air which is forced into the kiln. Air consists mainly of 78 volumes (76% by weight) of nitrogen and 21 volumes (23% by weight) of oxygen. Thus, it is necessary to introduce approximately 5 volumes of air for each volume of oxygen needed to obtain complete combustion. The nitrogen contained in the air does not enter the combustion process, only the oxygen reacts with the carbon, hydrogen, and sulfur to form the combustion gases.

Effect of Kiln Air on Combustion Efficiency.

It should now be apparent to the readers that control of the air supply for the kiln is as important as control of the fuel rate, because each is dependent on the other in their effect on combustion. In the following it will be shown that too little air (deficiency) as well as too much air (excess) is harmful to the economical operation of a rotary kiln.

When combustion is incomplete because of a deficiency of air, approximately 4500 Btu are released when one pound of carbon is burned to carbon monoxide. However, under conditions of complete combustion in burning the same amount of carbon to carbon dioxide, 14,500 Btu will be released, a difference of 10,000 Btu between complete and incomplete combustion, a clear indication of the importance of having at all times enough air available to accomplish complete combustion of the fuel. It should be pointed out that a rotary kiln, for safety reasons, is never allowed to be operated with such a deficiency in air supply. There is no room for any compromise in such a situation; the operator must do everything in his power to again obtain complete combustion of the fuel within seconds after the first indication of incomplete combustion is given. It is definitely the wrong procedure to increase the fuel rate at a time when a deficiency of air is already present and the burning zone starts to lose temperature.

More common is a rotary kiln operating with a large percentage of excess air, usually resulting from inefficient attendance of the kiln by the operator. The question might be raised now why too much air is also harmful to the efficient operation of a kiln, the reasoning being that with plenty of air available a total burning of the fuel will with certainty be secured. However, operating a kiln with a large percentage of excess air increases heat losses to the rear of the kiln and lowers the flame temperature. Furthermore, any excess air not required for combustion consumes valuable heat because this air too will be raised to the operating temperature of the kiln. In short, by consuming heat to raise the temperature of the excess air, less heat will be available for the actual burning process of the material in the kiln.

G. Martin,[3] in an extensive study of the thermodynamics of rotary kilns, has calculated that for every 1% of free oxygen present in the kiln exit gases, there is a loss of 0.4449 tons of fuel for every 100 tons of coal burned. That is, 0.4% of the heat introduced into the kiln by the fuel will be lost for each 1% free oxygen in the kiln exit gases. Consider, for example, a kiln producing 2040 tons clinker per day at an average specific consumption of 4.4 MBtu/ton clinker when the exit gas oxygen content is 1.5%. Burning the same amount of clinker per day but with 4.5% oxygen in the exit gas would result in a loss of 120 MBtu because of the excess air admitted to the kiln. At a common price of $1.80 per MBtu, roughly $67,000 per year would be lost.

Determining Excess or Deficiency of Air.

The foregoing discussion naturally leads to the question: How does an operator know when the kiln is operating with too much or too little air?

To assist the operator, analyzers are available that continuously determine the amount of oxygen and combustibles (carbon monoxide) in the kiln exit gases. Other analyzers, less frequently used, are available for determining carbon dioxide content. The Orsatt apparatus is used for making periodical analyses of gas samples to determine oxygen, carbon monoxide, and carbon dioxide content.

The percent oxygen contained in the kiln exit gases gives the best indication of the combustion condition in the kiln because this oxygen is directly related to the amount of air introduced and the amount of oxygen taken up during the process of combustion. We know that air contains 21% oxygen by volume. If there were no combustion reaction in the kiln, the same percentage of oxygen would be in the air leaving the kiln. However, because combustion reactions do take place in the kiln, most of the oxygen reacts with carbon, hydrogen, and sulfur to form the combustion products CO, CO_2, H_2O and SO_2. Thus, when no free oxygen can be found in the kiln exit gases no excess air has been introduced into the kiln.

The percent excess air can be calculated with the formula given by Perry, Chilton, and Kirkpatrick,[4] provided that no combustibles (CO) are present in the kiln exit gases:

$$\text{percent excess air} = \frac{100 O_2}{21 - O_2} \ K \qquad (5\text{-}6)$$

in which:
$K =$ 0.96 for bituminous coal
0.95 for oil
0.90 for natural gas, and
$O_2 =$ oxygen content of exit gases expressed as a percentage.

Fig. 5.1 shows this relationship for a typical rotary kiln.

The percent excess air can also be calculated from the results of a gas analysis obtained with the Orsatt apparatus. The following formula is especially helpful when a gas has a content of less than 1% oxygen because such gases usually contain traces of carbon monoxide also.

$$\text{percent excess air} = \frac{189(2 O_2 - CO)}{N_2 - 1.89(2 O_2 - CO)} \qquad (5\text{-}7)$$

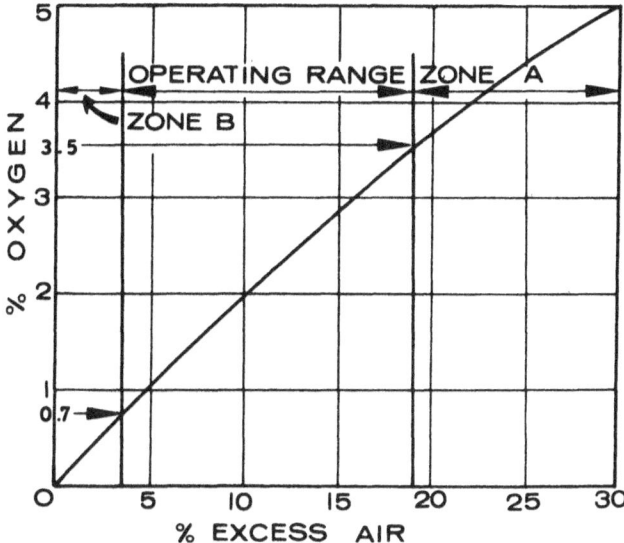

Fig. 5.1   Ideal operating conditions in the kiln occur when the kiln exit gas
contains between 0.7 and 3.5% oxygen.   Zone A indicates an excess of air,
resulting in excessive heat loss; Zone B indicates a deficiency of air, resulting
in the formation of carbon monoxide.

in which
    O_2 = percentage of oxygen
    CO   = percentage of carbon monoxide, and
    N_2 = percentage of nitrogen

Incomplete combustion is easily recognized when any percentage,
regardless of how small, is indicated on the combustible (CO) recorder. As
stated earlier, a rotary kiln is never allowed to operate under conditions of
incomplete combustion which means that at no time should there be any
carbon monoxide present in the kiln exit gases.

Now consider the question of just what constitutes the ideal operating
condition for efficient combustion and kiln control. At this point, theory
and practice are apt to take divergent paths. Many viewpoints and opin-
ions, some good, some in error, have been advanced, so it is well at this
point to review the theory of perfect combustion. First, most efficient
combustion takes place when there is neither carbon monoxide nor excess
air present in the kiln exit gases: that is, the oxygen and the combustibles

recorders should both have a reading of zero. Second, with any increase of either carbon monoxide or excess air, valuable heat is lost.

Application of this theory is of value only when perfect combustion conditions prevail within the kiln burning zone. In practice, however, this condition is rarely attained because many factors associated with design of the kiln work against such ideal conditions.

A typical example that spotlights this fact is the common observation that a small percentage of carbon monoxide can be found in the exit gases while at the same time there is also a small percentage of free oxygen present. This is in direct contradiction to the combustion theory that carbon monoxide shows only after all the excess air has been used up; that is, after the oxygen level has fallen to zero. In many instances the Orsatt analysis or the recorders have shown combustibles at oxygen readings in the range of 0.7%. This indicates that inefficient conditions are present when oxygen drops below 0.7%. The operator can observe the effects of changes in oxygen on the flame itself. If, for example, the kiln is operating at an oxygen content of 0.7% and the fuel rate is increased without an increase in the air flow in the kiln, such an action will cause a change in the color of the flame, the flame taking on a darker color at its outer rim, a sure sign that the flame temperature is dropping.

Through experience it has been found that a rotary kiln operates best when the kiln exit gases have an oxygen content of not less than 0.7% and not more than 3.5% under stable operating conditions. The optimum target point is between 1.0 and 1.5% oxygen. In addition, under no circumstances should there be any carbon monoxide present in the kiln exit gases. The given targets and ranges for oxygen levels do not apply at times when the kiln is in an upset condition.

New and more advanced types of gas analyzers have recently been installed in many kilns. These advanced technology analyzers are capable of detecting and recording minute traces of gas components such as O_2, CO_2, CO, as well as SO_2 and NO_x. Recordings on these units are usually in terms of ppm (parts per million) instead of percentages.

ppm	%
10	0.001
100	0.01
1,000	0.1
10,000	1.0

The thing to remember is that conventional gas recordings will only indicate the presence of CO when it exceeds approx. 500 ppm whereas the newer analyzers are capable of recording contents as low as 10 ppm. It is therefore necessary to reword the above-mentioned operating rule and state that immediate counteracting moves must be made whenever the kiln gases contain more than 100 ppm (0.01%) CO.

The fact that both too little and too much air for combustion is detrimental for fuel economy, has led to the practice in some plants of maintaining a constant fuel-to-air ratio at all times when the kiln is operating, even reaching the stage on some kilns in which controllers automatically adjust the total mass of air entering the kiln whenever a change in fuel rate is carried out. Although such an approach is fully justified from the viewpoint of combustion efficiency, it is not conducive to overall operating stability of the kiln.

Furnaces with less complex reactions than a rotary kiln can operate efficiently under the above principle. In the rotary kiln, however, there are many more factors that have a direct influence on the stability of the operation. For example, an important factor that is neglected when the principle of a constant fuel-to-air ratio is used to control a kiln, is the heat profile through the entire length of the kiln. A rigid application of this burning technique causes the temperature profile to change toward the back end of the kiln.

Table 5.1 shows the results of a burning technique in which the temperature profile has not been considered. Examples A and C follow the recommended practice of controlling the oxygen level within a range of 0.7–3.5%, and at the same time giving full consideration to the back-end temperature. Example A shows how the fuel rate and the air rate into the kiln are proportionally changed with succeeding increases in the kiln feed rate. With each feed rate increase there must be a corresponding increase in fuel rate and air flow rate as well as a rise in back-end temperature to take care of the increased production. This, then, is the procedure for stable kiln operation. Failure to make all these required increases could lead the kiln into an upset. There is little doubt that such an approach as shown in Example A will meet the approval of experienced operators.

Examples of upset kiln conditions are shown in B and C in which the feed rate remains constant. Example B shows the technique of maintaining a constant fuel-to-air ratio; that is, the oxygen percentage remains constant. This results in uniform combustion but very serious fluctuations in burning zone and back-end temperatures. Changes in back-end temperature

TABLE 5.1
TECHNIQUES OF COMBUSTION AND KILN CONTROL

Example A: Proportional increases in fuel rate, back-end temperature and I.D. fan speed (air rate) for increases in the feed rate.

Feed rate	Burning zone temperature	Fuel rate	I.D. fan speed	Oxygen	Back-end temperature
84	2800	180,000	380	2.5	1440
96	2800	200,000	395	2.5	1480
108	2800	220,000	410	2.5	1520
120	2800	240,000	425	2.5	1560

Example B: Fuel-to-air ratio constant (oxygen constant); feed rate constant, but kiln is in an upset.

Feed rate	Burning zone temperature	Fuel rate	I.D. fan speed	Oxygen	Back-end temperature
96	2600	240,000	425	2.5	1560
96	2800	200,00	395	2.5	1480
96	2950	180,000	380	2.5	1440

Example C: I.D. fan speed adjusted to compensate for temperature changes in the back end due to fuel rate changes. As in example B the feed rate is constant but the kiln is in an upset condition.

Feed rate	Burning zone temperature	Fuel rate	I.D. fan speed	Oxygen	Back-end temperature
96	2600	240,000	385	1.0	1480
96	2800	200,000	395	2.5	1480
96	2950	180,000	405	3.5	1480

result in uneven feed preparation (drying, calcination) in turn creating cycling operating conditons. Because the feed rate is constant, heat distribution should be unchanged to maintain the same drying and calcining conditons.

Example C is the recommended procedure for kiln upsets in which an attempt is made to maintain back-end temperature constant. Small adjustment are made to the induced draft fan speed to compensate for the temperature drops or rises that occur in the back end due to changes in the fuel rate.

The preceding discussion of combustion has centered on and is applicable to firing systems where the air supply to the furnace can be independently controlled. In other words, it applies to all kilns (wet, dry, semidry, preheater) that have one single-firing system in the lower part of the kiln (burning zone). It also applies to precalciner kilns that have tertiary air coming from the cooler for the flash furnace. The discussion on the proper air supply to the kiln and the optimum percent of excess air does not apply to precalciner kilns where all the air for the precalcining chamber (flash furnace) originates from and goes through the rotary kiln (i.e., no tertiary air duct present). In such kilns one single air-supply route must serve two combustion processes namely the flames in the burning zone and the flash calciner. Obviously, the excess air at the rotary kiln back end must be much higher otherwise incomplete combustion will take place in the flash furnace. Excess air from 60–90% at the feed end is not uncommon on such kilns. Through experience it has become known that large percentages of excess air renders a kiln inefficient by lowering the thermal level of the gases. As a whole, such a precalciner might be efficient but the fact remains that the burning of the fuel in the burning zone is done by large amounts of excess air hence no optimum conditions prevail here. This, too, is the reason why a precalciner, without tertiary air duct, usually does not achieve the high output rates that a precalciner of equal size with tertiary air duct does.

Combustion air requirements to burn fuel (solid or liquid) can be calculated as follows:

$$o = \left(1 + \frac{m}{100}\right)\left[0.11594A_C + 0.34783\left(A_H - \frac{A_O}{8}\right) + 0.04348A_S\right]$$

$$(5\text{-}8)$$

where:

o = combustion air required (kg air/kg fuel) — (lb air/lb fuel)

m = percent excess air in exit gas (use 5.0)

A_C = percent carbon in fuel

A_H = percent hydrogen in fuel

A_O = percent oxygen in fuel

A_S = percent sulfur in fuel

Products of combustion are mathematically determined by the following formulas:

$$CO_2 \text{ from fuel} = 0.0367A_CW_A \quad \ldots\ldots \quad (5\text{-}9)$$
$$SO_2 \text{ from fuel} = 0.02A_SW_A \quad \ldots\ldots \quad (5\text{-}10)$$
$$H_2O \text{ from fuel} = 0.09A_HW_A \quad \ldots\ldots \quad (5\text{-}11)$$

N_2 from fuel =

$$\left[\frac{A_N}{100} + 3.3478 \, (0.0267A_C + 0.01A_S + 0.08A_H - 0.01A_O)\right] W_A$$
$$\ldots\ldots \quad (5\text{-}12)$$

Subtotal $\qquad \ldots\ldots$

Add excess air: $\quad \dfrac{m}{100}$ (subtotal) $\qquad \ldots\ldots$

$G_t = $ total $\qquad \ldots\ldots \quad (5\text{-}13)$

where:

G_t = total combustion product (lb/lb clinker), (kg/kg clinker)
A_C = percent carbon in fuel
A_S = percent sulfur in fuel
A_H = percent hydrogen in fuel
A_N = percent nitrogen in fuel
A_O = percent oxygen in fuel
m = percent excess air in exit gas (use 5.0)
W_A = lb fuel fired/lb clinker or kg fuel/kg clinker

Carbon Dioxide in the Kiln Exit Gases.

The percent of oxygen in kiln exit gases is solely a function of the amount of air introduced into the kiln and the amount of air consumed in combustion of the fuel. In contrast to this, the percent of carbon dioxide in the exit gases is influenced by calcination of the raw feed in the kiln as well as by combustion.

When burning pulverized coal under perfect conditions, that is, when no excess air is present in the kiln gases, a maximum of about 18.3%

CO_2 would result from the combustion if there were no evolution of CO_2 from the raw feed in the kiln. For fuel oil this value is about 15.8%, and for natural gas, 12.2%. These percentages are obtainable only when perfect combustion conditions prevail in the kiln, which is very seldom the case. Any excess air present will cause a reduction in the percent of CO_2 in the gases. However, analysis of kiln exit gases in a cement plant shows a percentage between 22 and 28% CO_2. Therefore, a portion of the CO_2 in the kiln exit gases must originate from calcination of the raw feed.

Assuming that the CO_2 originating from the kiln feed remains constant, then any drop in CO_2 in the exit gases indicates an increase in air in the kiln. Knowing that a rise in excess air results in poorer fuel efficiency, it is then true that a decrease in CO_2 in the exit gases indicates a higher heat consumption for each ton of clinker burned. Hence the operator should try to obtain the maximum percentage of CO_2 possible during normal operation of a kiln.

If the only function of the CO_2 analyzer were to indicate fuel efficiency, there would be no reason for its use as this information is available from the oxygen analyzer. The conditions under which a carbon dioxide recorder becomes of value is when calcining conditions change in the kiln, that is, when the kiln is in an upset state. A change in the feed rate (including the dust return rate, if any), feed composition, or feed advancement within the kiln, can result in a change in the CO_2 content in the exit gases even at constant combustion conditions in the burning zone. The reason why the CO_2 changes in these instances is because more feed, less feed, or different feed, is being calcined in the kiln.

Changes that occur in the calcining zone cannot be seen by the kiln operator when he looks into the kiln to observe the burning zone because calcining takes place behind the burning zone. For this reason, a change in the character of the feed bed could occur unknown to the operator until it later becomes visible in the burning zone. Most common of such occurrences are kiln feed dust waves that flush at high speed into the burning zone, the arrival of large chunks of scale, or a change in the amount of feed entering the burning zone. By the use of a carbon dioxide analyzer and recorder, the operator is able to recognize such changes before they become visible in the burning zone, thus giving him time to make the necessary adjustments in the kiln control variables sooner than he could without the analyzer. If the operator notices an increase or decrease in the carbon dioxide recording, no adjustments having been made to such variables as fuel rate, air flow rate, or kiln speed, he knows that some change has taken place in the calcining zone. This is an early warning of possible changes in burnability that will soon occur. Once an operator has become accustomed to a particular kiln, he will know what settings in CO_2, O_2, fuel rate, feed rate, and other variables will most likely result in an efficient and stable kiln operation for a given feed composition..

A carbon dioxide analyzer is also helpful in obtaining stable operating conditions again after a kiln upset. Assuming that the kiln has been in an upset condition with oxygen lower and CO_2 higher than normal, then when the recorders start to show an increase in oxygen, with the corresponding decrease in CO_2 without any change in feed rate, the operator will know that calcining conditions behind the burning zone are improving and he can expect an increase in burning zone temperature within a few minutes.

Proper combustion requires that the kiln be operated in such a manner that:

1. The kiln exit gases have an oxygen content of not less than 0.7% nor more than 3.5% under normal operating conditions.
2. The kiln exit gases contain no carbon monoxide.
3. The kiln exit gases contain the maximum percentage of carbon dioxide.

In addition, the operator should:

4. Strive for optimum combustion conditions at all times by stabilizing kiln gas oxygen content between 1.0 and 1.5%.
5. Take immediate steps to eliminate combustibles in the exit gases whenever any carbon monoxide is detected therein.

In preheater and precalciner kilns it is of interest to note by how much (percent) the kiln feed is calcined after it leaves the preheater vessels and enters the rotary kiln proper. The laboratory checks this by analyzing the feed before and after the preheater tower, then calculates the result as follows:

$$\% \text{ calcination} = \frac{C-d}{C} \; 100 \qquad (5\text{-}14)$$

where:

C = ignition loss of fresh kiln feed

d = ignition loss of feed after precalcination

In the second group are the variables over which the kiln operator has control in order to obtain desired flame characteristics. These are:

Group 2

Fineness of the coal burned

Temperature of the fuel oil

Temperature of primary air

Temperature of secondary air

Flow rate of primary air

Flow rate of secondary air

Fuel rate

Burning zone wall temperature

Position of the burner

Degree of purity (dust concentration) of combustion air entering
 the kiln

Cross-sectional loading of the kiln

Air requirement of the fuel

Density of the primary stream (primary air plus fuel)

Density of the combustion gases

The nine italicized variables are the ones indicated by Gygi[*] as being the most important factors affecting the flame in a rotary kiln. A change in flame characteristics will take place whenever one of the variables in either group is changed. If the change results in a dangerous or undesirable flame, one or more of the other variables must be adjusted to counteract the bad flame condition.

Flame characteristics can vary considerably from one cement plant to another, sometimes even from one kiln to another in the same plant. The reason for this is that a flame must always be tailored to existing kiln designs and prevailing operating conditions. Clinker quality, refractory, presence of rings, and kiln equipment problems force an operator to obtain a certain flame that best fits the actual conditions in the particular kiln under consideration.

---

[*]Gygi, H. "Warmetechnische Untersuchungen des Drehofens zur Herstellung von Portlandzementklinker," Dissertation ETH, Zurich, 1937.

As a general rule in flame control, one should attempt to obtain the shortest possible fire and the highest possible flame temperature without adversely influencing clinker quality, coating formation, ring formation, and refractory life, or causing damage to the kiln equipment in the discharge area. Furthermore, the flame must not cause overheating in the burner hood, kiln discharge end, or cooler. Once the ideal flame characteristics have been obtained, the operator should make every effort to operate the kiln in such a fashion as to cause a minimum of disturbance to the flame. A flame should not willfully be changed during the course of kiln operation unless specific conditions necessitate a change.

Now consider in detail the several flame characteristics that were mentioned at the beginning of this chapter, and how these characteristics can be modified.

6.

The Flame

To an untrained eye, all flames in the burning zone of a rotary kiln appear more or less the same. As shown in Fig. 6.1, there is a short plume of air and fuel emanating from the nozzle, then at the end of the plume the fuel ignites and forms the flame. Flames vary considerably in length, shape, color, direction, and point of ignition. With time the novice learns to recognize these characteristics, their effect on kiln efficiency, and how to control them. Such terms as short, long, lazy, snappy, cold flame, and hot flame, the terms used to describe the different shapes and colors of flames, take on a real meaning and soon become commonplace in the language of the kiln operator.

The burner end of a kiln is equipped with a port through which the operator can view the burning zone. Because of the brilliancy of the incandescent fuel and gas, it is necessary to look through a special colored glass to protect the operator's eyes. When observing the flame, the operator must judge the point of ignition of the fuel, length, direction, and shape of the flame, and the color of the flame, the color being a good indication of flame temperature.

6.1 FLAME CHARACTERISTICS

Under stable operating conditions, no changes to the burner assembly or position are normally made unless hot kiln shell conditions make such an adjustment necessary. Therefore, the flame characteristics should theoretically remain unchanged. This, however, is not the case, because flame shape is influenced by many factors such as I.D. fan speed, secondary air temperature, primary air pressure and temperature, as well as changing conditions in the burning zone itself.

The factors that influence these characteristics can be subdivided into two groups. In the first group are the variables over which a kiln operator has little or no control. Some of these variables are impossible to change because they are an integral part of the kiln itself. Those that could possibly be adjusted, require either the help of a third party or a kiln shutdown. The following listed elements can be included in this group:

Group 1
Diameter of primary air nozzle
Orifice size of fuel burner
Diameter of the kiln
Design of the burner
Heat value of the fuel

6.2   LENGTH OF THE FLAME

When considering the length of a flame, one must make a clear distinction between two aspects of flame length. Flame length can be referred to as the distance between the nozzle of the burner and the end of the flame, or it can be expressed as the distance between the point where ignition of the fuel starts and where the reaction process of fuel combustion ends, that is, the length of the ignited part of the flame. The difference between these two concepts is often overlooked, but it is important to an understanding of kiln operation. Fig. 6.2 shows clearly the difference between these two measurements. Comparing the two flames A and B, it is seen that the distance between the nozzle of the burner and the end of the flame, 44 ft, is identical on both flames. If the flame length is considered as the ignited part of the flame only, then flame A is 40 ft long and flame B is 28 ft. In this book, the terms total flame length and ignited flame length will be used to distinguish the two measurements.

The variables having the greatest influence on flame length are the percentage of combustion air present and the velocity of the fuel-air mixture at the tip of the burner.

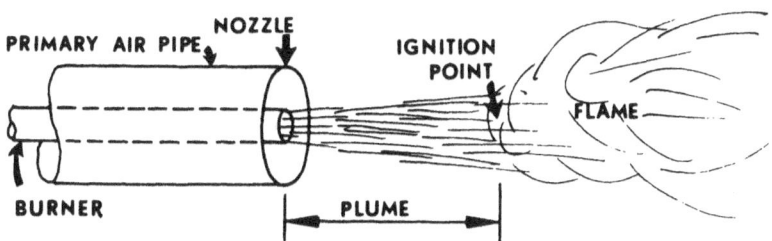

Fig. 6.1 The flames from all types of fuels have certain common characteristics.

It was seen in the previous section on combustion how a fuel, when introduced into the kiln, must combine with oxygen to start the chemical reaction of combustion. In other words, air which contains the oxygen, and fuel have to come into contact with each other to make the fuel burn. A lack of air can cause the flame to become too long, as the fuel then has to search for oxygen further back in the kiln.

Burner-tip velocity is **the** primary influence on flame configuration when coal and/or coke are fired. Unfortunately, direct-fired kilns most often demand a burner orifice size that is not to the best advantage of good flame control. On such systems, the burner size is more often governed by the coal mill's minimum air evacuation requirements and coal mill fan static pressure limitations. Flame control, unfortunately, is of secondary priority on such systems. Recently new, more sophisticated coal burner designs have come into use to eliminate some of the shortcomings of direct-fired systems but they still present problems in flame-control capability. It is known that a large percentage of indirect-fired kilns, particularly in Europe, operate with tip velocities of between 70–83 m/s (14,000–16,500 ft/min) and show exceptionally good flame patterns. However, direct-fired kilns more often are forced, by system design, to operate with velocities of between 45–66 m/s (9,000–13,000 ft/min) which mostly deliver a relatively lazy swirling flame. Many operators on such direct-fired kilns have unsuccessfully attempted to reduce the burner orifice size (i.e., to increase the tip velocity) and experienced coal evacuation problems from the mill. In other cases, the combination of higher tip velocity and relatively cold primary air have caused unstable and delayed ignition points in the flame. Coal burner designs on direct-fired kilns must be viewed as a compromise solution and is one area that leaves a lot of room for improvements.

Operators should be told by the engineering department what the typical tip velocity is on their kiln under various primary air pressures. As a rule of thumb, it takes a minimum of 35-m/s (7,000-ft/min) velocity in circular pipes to keep finely ground coal particles in suspension. Below this minimum velocity, coal could settle in the pipe and form pockets that could lead to dangerous backfires and/or explosions. Consult with the plant engineer about the minimum permissible, safe primary air pressure and once this level is established, **never introduce coal into the system unless you are absolutely sure that the air velocity and pressure are above this minimum permissible level.**

6.3   FLAME PROPAGATION SPEED

For coal-fired kilns, the primary air velocity should be at least twice as high as the flame propagation speed to prevent flashbacks of the flame. Flame propagation is usually considerably lower than the velocity needed to convey coal dust by means of primary air into the kiln. Therefore, the minimum velocity necessary to convey coal without settling in ducts takes precedence over flame propagation speed when setting air-flow rates or designing new burners.

W. Ruhland[6] in his investigation of flames, found that a decrease in nozzle diameter will at first give a shorter flame, but further decreases in diameter could lead to longer flames. He also found that no change in flame length will take place when the ratio between total combustion air and fuel introduced into the kiln remains constant. From this it can be concluded that the flame length is maintained unchanged when the percent oxygen in the exit gases remains the same. In other words, each time the fuel rate is changed, the rate of air going to the kiln should also be changed in order to maintain the same flame length.

The total mass of combustion air entering the kiln is the sum of the primary air, the secondary air, and the so-called parasite air which enters the kiln through leaks at the kiln discharge and burner hood. Because the operator has no control over the last air flow (unless he forgets to close one of the large doors of the burner hood), we shall consider only the first two mass flows.

The amount of total combustion air entering the kiln is governed mainly by the speed of the induced draft fan (usually called the I.D. fan). If

TABLE 6.1

FLAME PROPAGATION

a)   *(Metric Units)*

Primary air	Flame propagation (m/s)		
m³/kg coal	30% VM*\n5% ash	30% VM\n15% ash	10% VM\n7% ash
1	4.0	3.9	2.2
2	8.8	7.5	3.8
3	13.0	11.0	5.4
4	14.4	11.7	6.7
5	14.0	11.1	6.3
6	13.1	10.0	5.5
7	12.2	9.0	4.9
8	11.4	8.1	4.5
9	10.8	7.5	4.1
10	10.3	6.9	3.8
11	9.8	6.6	3.6

b)   *(English Units)*

Primary air	Flame propagation (ft/min)		
ft³/lb coal	30% VM\n5% ash	30% VM\n15% ash	10% VM\n7% ash
20	950	900	500
40	2250	1900	900
60	2800	2300	1300
80	2750	2180	1240
100	2550	1900	1050
120	2320	1670	930
140	2150	1500	840
160	2020	1350	750
180	1900	1250	690

* volatile matter

the fan speed is increased with the fuel rate constant, more combustion air enters the kiln. Conversely, a decrease in I.D. fan speed will result in a decrease in total combustion air entering the kiln. The operator has at his disposal a quick indication with the kiln exit gas oxygen analyzer to determine if the flame theoretically is getting longer or shorter. Lack of combustion air in the kiln is indicated when the oxygen analyzer reading approaches zero and the carbon monoxide combustible instrument shows that carbon monoxide is present in the kiln exit gases.

To explain this effect on flame length consider several examples.

Case 1:   Oxygen reading is 2.0%, I.D. fan speed is increased such that the oxygen reaches a value of 4.0%. Result: total flame has been lengthened.

Case 2:   Oxygen reading is 2.0%, I.D. fan speed is decreased such that oxygen decreases to a value of 0.7%. Result: total flame length has been shortened.

Case 3:   Oxygen reading is 0.7%, I.D. fan speed is decreased such that the oxygen obtains a value of 0.1% and a small amount of CO is also indicated. Result: total flame has been lengthened.

Case 4:   Oxygen reading is 0.1% and combustibles are showing. I.D. fan speed is increased such that the oxygen increases to 0.7% and no more combustibles are indicated. Result: total flame has been shortened.

From these examples it becomes clear that it is not possible to state simply that an increase in air flow will result in a lengthening of the flame or a decrease will shorten the flame, but that the resulting changes are dependent on the percent oxygen present before the changes in air-flow rates are made.

The total flame length could become infinitely long under the condition of an extreme lack of air. As a matter of fact the fuel could travel the entire length of the kiln without igniting because not enough air is present in the kiln itself. Upon entering the dust collector at the kiln rear where plenty of air is available, the fuel could explode if the temperature is high enough to ignite the fuel. One can easily visualize the results of such a disastrous occurrence.

A word of caution must be expressed at this point. Considering how flame length can be changed by changes in the I.D. fan speed, and knowing the desirability of maintaining the flame length constant, could lead one to the conclusion that the I.D. fan speed should be proportionally changed whenever a change in the fuel rate is carried out. Later on in this book, the pitfalls of such an approach because of its effect on back-end temperature and overall kiln operation in general will be pointed out. One must at all times correlate the different control functions together for the entire kiln operation and take the approaches that will be the most beneficial for stability of kiln operation. For example, maintaining the oxygen content at a fairly constant level while varying the kiln speed is one technique of kiln burning that is not recommended for prolonged stable kiln operation.

In the technique of kiln burning stressed in this handbook, the operator gives special attention to the burning zone temperature and back-end temperature, at the same time maintaining the oxygen content in the kiln exit gases within a range of not less than 0.7% and not more than 3.5%. This "floating" control of the oxygen results in changes of the total flame length. Experience has shown that the flame length changes are of minor magnitude when the oxygen level is controlled within this range, and far more benefits can be derived in operating stability than when the oxygen would be strictly controlled at a constant level.

6.4   IGNITION OF THE FUEL

The initial ignition (start of combustion) of the fuel is primarily dependent on sufficient heat to ignite the fuel and on sufficient air to obtain combustion of the fuel. A deficiency in either one of the two in the area where the fuel leaves the burner and enters the kiln can cause a delayed ignition of the fuel.

During the initial period of a kiln start, the kiln operator may experience difficulties in obtaining or maintaining continuous ignition of the fuel. This is understandable because during such periods furnace temperatures are usually too low to ignite the fuel properly. One must resort to the help of a pilot burner (auxiliary torch) placed at the mouth of the burner pipe to obtain good ignition. It is a good practice to leave the pilot burner in this position until the kiln interior has acquired the necessary heat to sustain continuous ignition. Operator jargon often refers to this as the time when the wall temperature in the burning zone can support the flame.

Once the kiln has reached operating temperature and continuous burning of the fuel has been obtained, the ignition point of the flame can be willfully adjusted or inadvertently changed by changes of certain operating variables.

A decrease either in primary or secondary air temperature can move the ignition point further into the kiln. Design of the kiln burner hood and the burner also plays a part in the point of ignition of the fuel. For example, the secondary air stream could enter the kiln in such a fashion that a rapid contact with the fuel is not possible. The fuel will thus ignite at a point further in the burning zone, or a burner can be designed so that a rapid mixing of the air with the fuel is promoted which results in an earlier ignition of the fuel once it enters the kiln.

It must be pointed out that no clear-cut general answer can be given to the question of where the point of fuel ignition shoud be located in a rotary kiln. This depends on the type of fuel used and the overall conditions that result from a given flame structure and ignition point location. Although it is advantageous to have the fuel ignite as early as possible after it leaves the burner, this can in many cases create overheated conditions in the kiln nose area and the burner hood, as well as in the cooler. In general, one can say that the earliest possible ignition point of the fuel is desirable under the condition that there will be no harmful effects on kiln equipment as a result.

The plume, as shown in Fig. 6.1, is the part of the flame between the burner nozzle and the point of ignition of the fuel. On coal-and-oil-fired kiln this plume is recognizable as a black cloud or jet. On coal-fired kilns, a shorter plume (earlier ignition) can be obtained by grinding the coal to a larger surface area; that is, grinding the coal finer. Earlier ignition on oil-fired kilns can be accomplished by increasing the temperature of the fuel oil, and by selecting a smaller oil burner opening (orifice, or tip size) such that the oil will be better atomized. For both types of fuels, shorter plumes can be obtained by designing the burner so that a rapid mixing of air and fuel takes place at the burner tip. Control of plume length in a gas-fired kiln is not easily accomplished, but is not particularly important. Kiln interior temperature, primary air temperature, and secondary air temperature, three very important factors affecting plume length, are the variables that are most frequently apt to change during operation of the kiln. An operator will always try to hold these temperatures within close limits, because a large change in any one can lead the kiln into an upset condition and affect overall flame characteristics in an undesirable manner.

At this point, a word of caution must be inserted with respect to the primary air temperature. On many rotary kilns, preheated primary air is used to promote ignition of the fuel and to form a specific desirable flame. The ignition temperature for various fuels commonly used on rotary kilns can be as low as 800 F (425 C). For safety reasons, temperatures within the primary air duct must therefore be kept well below this critical level at all times. This becomes especially important on coal-fired kilns where the coal dust is mixed with the primary air inside the burner pipe. Exceeding this critical temperature could ignite the coal inside the burner pipe with possible disastrous results. The most probable times for this adverse possibility are the periods immediately after a fire has been cut off or lighted during short kiln shutdowns; that is, when the burning zone is still at an elevated temperature. These high burning-zone temperatures could easily find their way into the primary air duct. Explosions have occurred in the primary air pipe because fuel was able to accumulate inside the pipe during a shutdown as a result of a leak in the coal feeder mechanism or leak of the fuel shut-off valve. Because of the potential danger created by such mechanical deficiencies, these fuel valves and feeders should be inspected frequently on a routine basis and any malfunction repaired at the earliest possible time. To safeguard against an explosion due to ill-adjusted air-to-fuel ratio during the initial moment of firing a kiln, an operator will always make sure that enough air is present in the kiln for the intended fuel rate and that enough heat is present to ignite the fuel *before* the fuel is introduced into the kiln.

So far, we have centered our attention on the ignition point and the plume of the flame, giving us some insight into one important aspect of ignition. Now let us consider another important element, the aspect of total ignition of all the fuel introduced into the kiln.

It is absolutely essential that all the combustion reaction (burning of the fuel) takes place in the atmosphere of the burning zone. Impingement of partially burned fuel upon the feed bed or kiln wall must be avoided at all times when a kiln is being fired with coal or oil. Impingement of unburned fuel particles on the wall and feed bed is not only harmful to the refractory and clinker but also acts as a bad influence on fuel efficiency and kiln operating stability. Alignment of the burner, fineness of the coal, and the velocity with which the fuel and the primary air enter the kiln are the most important factors governing impingement. Fineness of the coal must be such that all coal particles are burned completely while they are in suspension in the kiln atmosphere. The same holds true for fuel oil firing;

oil droplets must be small enough so they burn completely in the same manner. It is obvious that the velocity of the fuel as it enters the kiln must be adjusted properly so the fuel particles have enough velocity to remain in suspension while they burn, but not so high that they are carried too far into the kiln. A wrong selection of primary air pipe nozzle diameter or orifice size of oil and gas burner nozzle could result in either insufficient or excessive velocity of the fuel-primary air mixture.

All coal-mill systems are equipped with thermocouples to measure the so-called coal-mill outlet temperature. The maximum permissible and recommended temperature is 74 C (165 F) for high-volatile coal and 85 C (185 F) for low-volatile coal. This is considerably below the ignition temperature for these types of coal but considered essential as a safety measure to prevent premature ignition in the mill or primary air pipe.

There is another control function an operator has to keep close watch over, that is, the coal-mill inlet temperature. The coal mill's function is not only to grind and pulverize but also to dry the coal. This is accomplished by using hot excess air from the clinker cooler, cooler air which can fluctuate quite frequently from a high of 650 C (1202 F) to a low of 200 C (393 F). From the previous discussion of the combustion triangle (Fuel—Air—Heat), it can be seen that such high air temperatures would be sufficient to start a fire in the coal mill if this hot air temperature is allowed to fluctuate unchecked. For this reason, the hot air duct to the coal mill is equipped with a safety ambient-air inlet device that allows, by means of a damper, the mixture of hot with cold air, and thus controls the coal-mill temperature within safe limits. Coal, too, fed to the mill for grinding and drying, can fluctuate considerably in its requirement for heat depending on its moisture content and the amount being fed to the mill. Therefore, it becomes apparent that here three highly varying entities combine together to form one of the seemingly more complicated control functions on the kiln system:

changing cooler air temperatures
changing coal moistures
changing coal feed rates

This process control function is handled with a few simple instruments. First, the designated main control (input) variable is the coal-mill **outlet** temperature from a thermocouple located near the point where the coal/air mixture leaves the mill. The operator regulates the setpoint for this tem-

perature on the automatic controller, which in turn will almost continuously adjust the ambient air-tempering damper in the hot inlet-air duct
to maintain the desired mill outlet temperature at this predetermined setpoint. In the case where the cooler excess air is so hot that the ambient air
damper, even if wide open, is not sufficient to hold the coal mill outlet
temperature below safe levels, there is another control loop which can be
found on most coal mill systems. Here the input variable is the coal mill
inlet temperature. Here, too, the operator sets a predetermined setpoint
for this temperature on a separate controller which in turn drives a damper
that allows for ambient air mixing at a second location (usually at the
beginning of the hot-air duct near the cooler). The newer, more modern
sytems also incorporate automatic CO_2 injection as an added safety device
to prevent coal-mill fires and/or explosions. If properly designed, operated,
and maintained, these systems are safe. There are, however, some important aspects and considerations an operator must always keep in mind
regardless of how well or calm the system might appear. It is important
to regard it with caution and expect the unexpected. Here are a few
pointers:

a) Both of the above-mentioned controllers must be set so that they
   respond very rapidly to a slight change in temperature, i.e., the
   tempering damper adjustment must be almost instantaneous
   whenever a slight deviation from setpoint occurs.
b) All input-output devices must be properly maintained, they must
   function, and must be immediately attended to if adjustments are
   needed. No shortcuts can be taken.
c) Under normal operating conditions, both ambient-air-tempering
   dampers must be set in a position so that there is plenty of lee-
   way (adjustment capability) left for the damper to move in either
   direction. Case in point: it wouldn't do a controller any good if
   the tempering damper, under normal operating conditions, is always in the fully open position.
d) It is important to communicate with the plant engineer so as to
   get a clear understanding of the maximum and minimum permissible temperatures and pressures. A record of these temperatures should be kept in a standard operating-procedure notebook.
e) Learning the coal-mill system should be the first priority in the
   learning program. Becoming fully familiar with all its aspects

and learning what to do when things go wrong are essential. Just
learning what buttons to push for what isn't enough.

Having started this discussion with flame control and arrived at a discussion of where hot air is leaving the cooler, demonstrates how inter-
related the functions of a rotary cement kiln are. A change in one variable
can ultimately cause a change at another location that, on the surface,
doesn't appear to be related at all.

Returning to the discussion of flame control, the theoretical flame temperature can be approximated by the following formula:

English or Metric units:

$$T = \frac{H_v}{1.11As} \tag{6-1}$$

where:

T = theoretical flame temperature (°C or °F)
A = combustion air required (lb/lb or kg/kg of fuel)
H_v = heating value of fuel (kcal/kg or Btu/lb)
s = specific heat of combustion gas (use 0.29)

The values obtained by these calculations are usually in the magnitude of
3700–4800 F (2040–2650 C). At these temperatures the flame should
have a dazzling white color. Comparing this with the color usually observed in a rotary kiln, it becomes quite obvious that actual flame temperatures fall quite short of these calculated maximum possible temperatures.
Flames with a light yellow to white color [2400–2800 F (1320–1540 C)]
and yellow to light yellow [2000–2400 F (1090–1320 C)] are common in
rotary kilns. The flame colors discussed here are actual flame color and not
the color an operator sees through a filter glass.

Operators often speak of hot and cold flames. These terms, although
contradictory because there is no such thing as a "cold" flame, are fully acceptable because they state in unmistakable terms a given characteristic of
a flame. What the operators mean by these expressions are high and low
flame temperatures.

6.5   OXYGEN ENRICHMENT

The steel industry has made some large improvements in regard to flame temperatures in open-hearth furnaces by oxygen enrichment of the combustion air, resulting in much higher flame temperatures, which in turn improves the efficiency of this type of furnace. Oxygen enrichment of the combustion gases in a rotary cement kiln has been attempted in a number of cement mills, but because of the high cost the process has been abandoned. Using oxygen enrichment could create an undesirable result because of the extremely intense, hot, and short fire that is characteristic of this type of flame. Extreme high temperatures might be localized close to the discharge area of the kiln which could cause damage to the refractory as well as to the burner hood and cooler.

It is quite conceivable that oxygen enrichment might, once again, find some favor and acceptance in a precalciner kiln. To the author's knowledge no plant has tried this so far. As previously stated, there are two types of precalciner kilns: the ones with tertiary air ducts and those that draw the combustion air for the flash furnace from the kiln itself. It has been mentioned that the latter must operate at very high kiln exit-gas oxygen levels to secure the proper air supply to the flash furnace. This leads to lower flame temperatures in the kiln itself due to the excess air present but makes this type of kiln considerably less costly than a precalciner with a tertiary air duct. Besides, control of the kiln overall is also made easier by not having to control the air flow and temperature through the tertiary duct. Oxygen enrichment at the flash furnace could possibly allow such kilns, without a tertiary air duct, to operate the kiln at normal oxygen levels in the kiln exit gas of say 0.8–1.5%. This idea would require research before deciding whether or not it could be feasibly implemented.

Flame D, Fig. 6.2, shows an extreme condition in which part of the flame impacts against a clinker ring, a common condition on some kilns. Although such a ring formation speeds up the reaction of combustion (if enough air is present), this condition is extremely harmful to the coating and refractory in the area immediately preceding the ring. It is not uncommon, on kilns that experience such frequent ring problems, to find the shortest refractory life of the entire burning zone lining in this particular area where flame impingement upon the wall takes place. This is one of the reasons why it is a good practice to remove (shoot) the ring before it causes flame impingement, that is, before the ring has grown too large.

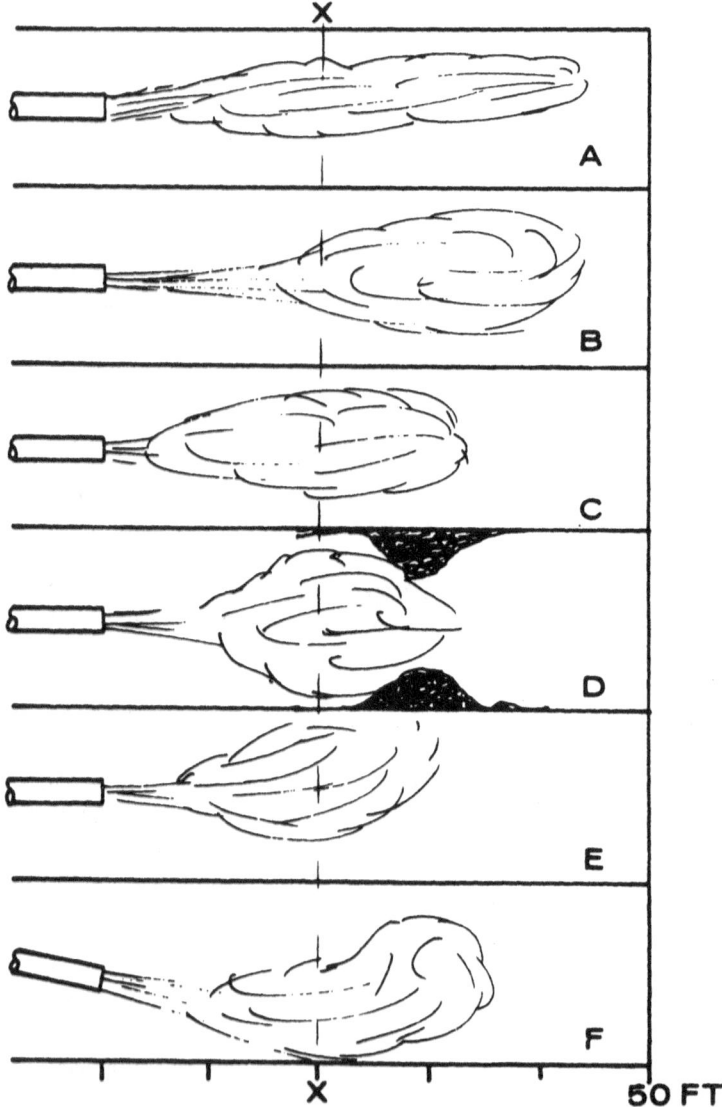

50 FT

Fig. 6.2   In an exaggerated and simplified form, the illustrated flames show the essential differences between the several flames.

6.6 SHAPE OF THE FLAME

As previously pointed out, it is desirable to operate a rotary kiln with the total flame length as short as possible. A short flame extends the length of the calcining zone, which can often lead to increased production capabilities of the kiln.

Now consider Flames A and C in Fig. 6.2. Although the same amount of fuel is being burned in both flames, the two flames are entirely different. Flame A represents what operators call a "lazy" fire because heat from this flame is released over a relatively long distance of the burning zone. Flame C is a "snappy" fire in which the heat is released over a shorter length of burning zone. Both types of flames can be found in rotary kilns, but flame C is more favorable for ease of kiln control, clinker quality, and fuel efficiency.

Flames such as shown in example A usually have a tendency to float about the kiln atmosphere in an unstable fashion, but flame C, with its so-called better body, tends to remain fixed in position better when small changes in air-flow rates or temperature occur. Flame C, with only a few exemptions, is the most favorably shaped flame for a rotary kiln. The burning fuel (flame) expands in a uniform wedge equally in all directions along the kiln axis, and is not too short and not too long in total length. Stability of the flame must always be maintained and stable flame characteristics (when the flame is favorably formed) pay off with long refractory life and good kiln control conditions.

6.7 DIRECTION OF THE FLAME

One of the most important things for the kiln operator to remember is the fact that the path of the flame is not a straight line, and it does not remain constant at all times. Thus, when the primary air pipe is set for the fuel-primary air mixture to travel toward a certain target in the burning zone, this will not guarantee that the flame itself will follow this particular path for its entire length. The flame has a tendency to lift upward toward the top of the kiln arch, because of the buoyancy resulting from uneven entrance of secondary air into the kiln. This hot secondary air stream enters the kiln from the bottom, and moves upward in a curved path

toward the axis of the flame a few feet past the burner nozzle, where it
starts to mix with the fuel-primary air stream.

Another factor influencing flame direction is the mechanical condition
of the primary air pipe nozzle. Because of the high temperatures prevailing under normal operating conditions in the burner hood area, this
primary air pipe nozzle is especially vulnerable to warping and distortion,
resulting in uneven envelopment of primary air around the fuel jet causing
erratic flame shapes and direction.

6.8   ADJUSTMENT OF FLAME DIRECTION

On some rotary kilns, the primary air pipe is fixed slightly below the
center of the kiln, in an effort to offset the flame buoyancy. Most kilns,
however, have facilities for adjusting the position of the primary air pipe
and the burners, thus the operator can change flame direction and, to a
limited degree, change the shape of the flame. Every adjustment in the primary air pipe or burner position will bring about a change in flame characteristics and can therefore affect burning conditions in the burning zone.
Buoyancy of the flame can usually be counteracted by a slight downward
tilt of the primary air pipe so the main body of the flame will be centered
at a certain distance inside the burning zone. In Fig. 6.2, flame E shows a
buoyant flame with the primary air pipe centered in the kiln axis. By
tilting the pipe down slightly, the flame assumes the position shown by
flame F.

Because so many factors influence flame direction, it is necessary to fix
the primary air pipe position according to the actual flame condition encountered in each individual kiln. To do this, the operator will observe the
flame in the kiln, paying particular attention to the position of the flame
body at point "x" in Fig. 6.2. Fig. 6.3 is a cross section of the kiln at
"x."

The centerpoint of the flame at this location can be in any one of the
indicated squares depending on how well the flame direction is adjusted. It
is generally assumed that the best heat exchange between the flame and the
feed takes place when the flame is pointed slightly toward the feed bed,
hence the most favorable target for the flame is position 2A or on the kiln
axis (2B). On the other hand, if the flame is directed too close to the feed
bed (3A), there is danger that part of unburned fuel (especially on coal and

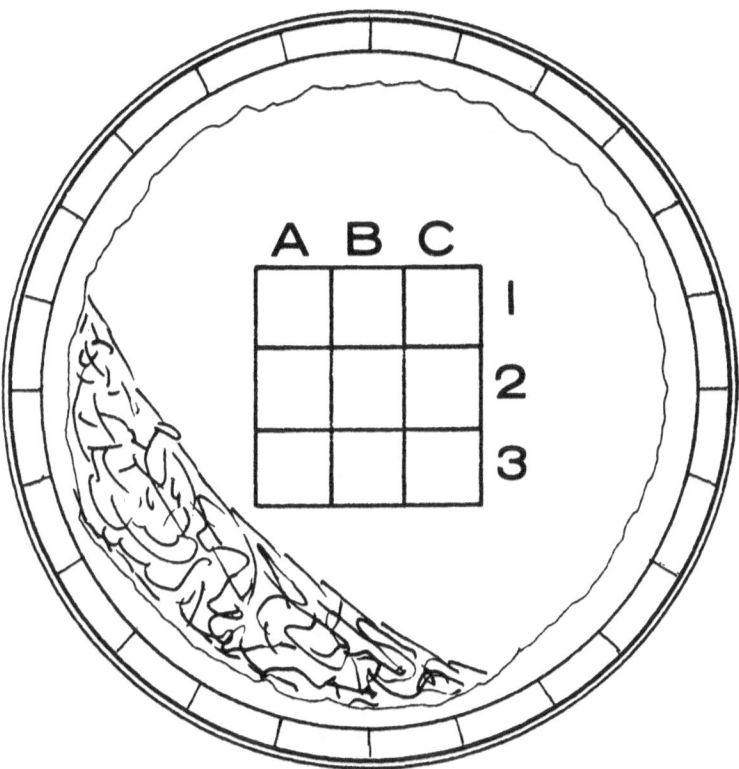

Fig. 6.3   Imaginary targets in the kiln indicate the correct and incorrect areas toward which the flame should be directed.

oil fires) could enter the feed bed, a condition that is highly undesirable. A flame target leaning too much toward the kiln wall (1C, 2C, or 1B) could result in flame impingement upon the coating, an action which is known to shorten the life of the refractory.

It has been found that the best method to fix the burner position is in the following manner: Prior to firing up a kiln (after a relining job), measure a given distance, e.g., 20 m (65 ft) from the kiln discharge and lay a white rag at this location. Then, go back to the firing floor and look through the burner (a cross-hair at the tip is helpful) to target the burner direction toward this rag. Once this is accomplished, tilt the burner slightly toward the load-side, e.g., the 5- or 7-o'clock position from the rag). The distance to place this rag is found by trial and error, for in some instances 80 ft is required and other times only 50 ft is necessary to obtain optimum flame positioning. Once this position has been found, however, it should not be changed and should become standard procedure on all subsequent kiln-start preparations.

Although no clear-cut standards can be given for the direction of the flame for all kilns, because each kiln has this position specifically tailored to its own particular problems, design, or conditon, there are, however, a few rules that can be applied to all flames, regardless of what rotary kiln is under consideration. These are:

a) When the primary air pipe nozzle has accidentally been warped, resulting in an erratic flame shape and direction, immediate steps should be taken to repair this condition.

b) A flame should never be allowed to impinge upon the coating or bare refractory for a prolonged length of time.

c) A flame should never be allowed to strike too hard upon the feed bed.

d) Oil burners or gas burners should be centered well in the primary air pipe in order that an even envelopment of air around the fuel jet takes place.

e) Flame direction should be adjusted only when the kiln is in stable operating condition and the temperatures, fuel pressures, and air-flow rates are at a normal level. Flame direction changes can be caused by unusual operating conditions. If an attempt were made to adjust the flame direction at such times, there will most likely be an undesirable flame once the kiln returns to normal operating conditions again.

f) It is better to make the desired adjustments in flame direction in several small steps instead of a large one in order that the operating stability of the kiln is not affected adversely.

g) Once the ideal flame direction has been obtained, the primary air pipe position should not be changed unless a definite reason (such as to combat ring formation or hot shell conditions) makes it desirable.

h) To protect the primary air pipe from possible damage during a shutdown a certain amount of primary air flow must be maintained until the temperature inside the kiln is low enough (approximately 600 F or 315 C) that the pipe cannot be damaged. Upon power failure when the primary air fan stops, the primary air pipe must be immediately removed from the burner hood.

6.9 TEMPERATURE OF THE FLAME

The amount of heat released by the flame depends on the flame temperature, and the flame temperature indicated by the color of the flame (Table 6.2). By observing the color, the operator can estimate the flame temperature within certain rather broad limits. However, before discussing color of the flame, it is first necessary to point out that an increase in fuel rate does not always increase heat output of the flame, nor does a fuel rate decrease always reduce the heat output. The reason is that combustion of the fuel is also dependent on the proper amount of air (oxygen) available and upon the prevailing temperatures of the gases and the kiln wall in the burning zone. Liberation of heat by the fuel, because of its dependence upon these factors, can at times be somewhat erratic and confusing to the newcomer who is learning to operate a kiln.

A prime example, and because of its simplicity and frequency of occurrence, the best worth mentioning, is the condition in which the kiln is operating with a very low content of oxygen in the exit gases, e.g., 0.5%. The operator notices that the burning zone temperature starts to decrease. The wrong procedure in many such instances is to increase the fuel rate, hoping that this will help to raise the temperature. The contrary is most likely to happen; that is, a further decrease in flame temperature, because of the insufficient air supply available for good combustion of the fuel. The proper procedure in such an instance would be to reduce the kiln speed to allow more time for the clinker to burn properly, or to reduce the fuel rate to improve combustion of the fuel. A third possibility is to increase the primary air flow to compensate for the lack of necessary combustion air brought about by the increase in fuel rate.

The same or similar reaction can take place with a sudden large drop in secondary air temperature, or when too much air (large percentage of excess air) is introduced into the kiln by excessively high I.D. fan speed. Both of these occurrences can result in a decrease in flame temperature.

Under any operating condition, regardless of the fuel rate, it is desirable to obtain the highest possible flame temperature. In other words, the best fires in a rotary kiln are the ones with a very bright color. Orange and red-colored flames are undesirable because they possess a lower flame temperature. The values shown in Table 6.2 are not only helpful in respect to flame temperatures, but can also be of service to the operator for any kind of observation in the kiln system where elevated temperatures prevail such as the burning zone wall, feed bed, and kiln shell.

TABLE 6.2

CORRESPONDING TEMPERATURES
FOR OBSERVED COLORS

Color	°F	°C
Lowest visible red	875	475
Lowest visible red to dark red	875–1200	475–650
Dark red to cherry red	1200–1375	650–750
Cherry red to bright cherry red	1375–1500	750–825
Bright cherry red to orange	1500–1650	825–900
Orange to yellow	1650–2000	900–1090
Yellow to light yellow	2000–2400	1090–1320
Light yellow to white	2400–2800	1320–1540
White to dazzling white	over 2800	over 1540

Of specific interest for the kiln operator are the following factors that serve to raise flame temperature:

a) Increasing the secondary air temperature.
b) Using less primary air, thus making it possible to utilize more secondary air which is preheated to higher temperature.
c) Promoting rapid mixing of the air and fuel upon leaving the burner by improving the design of primary air pipe and burner.
d) Better atomization of the fuel oil by increasing the fuel oil temperature and selecting a smaller burner orifice size (tip size), or employing a mechanical device in the burner nozzle to bring about better atomization.
e) Operating a kiln with neither a deficiency nor excess of air by maintaining oxygen content of not less than 0.7% and not more than 3.5%.

7.

Heat Transfer

In this chapter a review of the fundamental concepts of heat transfer is presented. The second law of thermodynamics states that heat always flows from a point of higher temperature to the point of lower temperature. Heat can be transferred by *radiation*, which is the transfer of heat from one body to another by means of heat waves without the two bodies being in actual contact with each other. Another type of heat transfer is by *conduction*, where the vibration of molecules of solid bodies collide with adjacent molecules and thus transfer energy between each other while in contact. The third type of heat transfer is known as *convection* which occurs in gaseous and liquid fluids. In this type of heat transfer, the molecules, after having collided with each other, are free to move and circulate and thus are able to transmit more energy to other molecules having different temperatures.

In a rotary kiln, the largest amount of heat transfer occurs by radiation and conduction whereas heat transfer by convection occurs only in small amounts. Although there are many more occurrences of heat transfer, the following examples explain some of the more prominent heat exchanges that take place in a rotary kiln:

RADIATION:	Flame ——————————————→	feed bed
	Hot kiln wall ———————————→	feed bed
CONDUCTION:	Heat from kiln interior ————————→	kiln shell
	Kiln chains ————————————→	feed bed
CONVECTION (in heat exchangers):		
	Super-heated steam———————→	oil in tubular pipes
	Hot kiln exit gases ————→	steam in boiler tubes

83

Kiln burning can be defined as exercising control over all heat transfers that take place within and in the proximity of the rotary kiln. A kiln operator has little control over the *type* of heat exchange that takes place. Whether it will be radiation, conduction, or convection is primarily fixed and governed by the design of the equipment. But control over the *amount* of heat transfer that takes place is posible since the operator's primary function is to control the temperatures at multiple locations of the kiln system. Temperature and heat transfer are quantities that are dependent on each other and inseparable.

7.1   THE THERMAL WORK REQUIRED IN A CEMENT KILN

Regardless of what type of kiln is used, there is a fixed amount of thermal work that has to be done within the system to obtain what is known as cement clinker. This thermal process is accomplished in several stages, stages which are discussed in more detail in a later chapter. The key thing to remember is that there must be a minimum amount of heat (temperature) imparted into the feed so that these phases can progress properly. W.L. De Keyser[1], based on laboratory investigations, has made determinations as to the temperatures at which these phases occur. Likewise, G. Martin[2] has done extensive studies of the thermodynamics of a rotary kiln, however, he based his investigations on practical kiln data. From these two works the following table has been compiled which informs in very broad terms what actually goes on inside a rotary kiln.

The concept of low-grade and high-grade heat, first advanced by G. Martin, is a very important aspect of kiln operation. It simply means that regardless of how much heat is applied to the rear of the kiln, if that heat is below 805 C (1481 F) there can be no other thermal work done other than drying the feed and driving off the chemically combined water from the clay minerals. No calcination can take place. The same holds true in the burning zone. Unless a minimum of 1400 C (2552 F) is reached for the feed in the burning zone, there will simply be no clinker produced that could meet the quality standards.

7.2   HEAT PROFILE

Temperatures differ throughout the kiln and a very important factor is the temperature difference between the material and the gases at any given point in the kiln. A recorder chart of the gas temperature shows gas tem-

TABLE 7.1

PHASE FORMATIONS

	Thermal work	Minimum temp.		
		°C	°F	
1.	Evaporation of free water from the feed	100	212	Required low-grade heat 363 kcal/kg (652 Btu/lb)
2.	Evolution of the chemically combined water in the clay minerals	550	1022	
3.	Evolution of CO_2 from calcium and magnesium carbonate (calcination)	805	1481	
4.	Formation of interm phase C_2F	800	1472	
5.	Formation of interim phase CF	900	1652	
6.	Formation of interim phase $C_2S + C_2AS + CF + C_2F$	1000	1832	
7.	Formation of interim phase $C_2S + C_5A_3 + C_5A + C_3S + CF + C_2F$	1100	2012	Required high-heat grade heat 511 kcal/kg (919 Btu/lb)
8.	Formation of interim phase $C_2S + C_3A + C_5A_3 + C_2F + C_3S$	1200	2192	
9.	Formation of C_3A, C_3S, C_2S, C_2F	1300	2372	
10.	Equilibrium in C_3A, C_3S, $C_2S + C_6A_2F$	1400	2552	

perature only, and this temperature never corresponds to the temperature of the material in the same location. Also, conditions could change in such a way that the gas temperatures would show a considerable change but the material temperature remains unchanged for several minutes after. This applies to the feed-bed behavior within the rotary kiln itself since the rotary cylinder in itself is a relatively poor heat exchanger.

7.3 HEATING THE FEED

Radiation of heat from the hot gases to the feed bed occurs only at the surface of the bed that is exposed to the gas. Thus the temperature of the bed is lowest in the center and highest on the surface. In the burning zone the kiln feed, being now in a rather sticky condition, is in a constant state of agitation as, aided by the rough and uneven kiln coating, it rises along the upward-moving side of the kiln, then tumbles back. Because of this tumbling action, the surface layer is constantly being folded back into the mass of the feed, where the hot particles then transfer heat to the colder particles by conduction. Meanwhile, new particles are being exposed to radiation from the hot gas and the process is continuously repeated. In the calcining zone and toward the back end of the kiln, where the surface of the lining is much smoother and the feed still in a more or less fluid state, the bed is turned over very little, as it slides in a zig-zag course down the kiln. The bed is first lifted up the kiln wall, then when it reaches a certain height it slides down and forward without turning over to any extent. However, the turnover is larger in kilns with pelletized feed, such as wet or semiwet process kilns.

Heat exchangers, in the form of chains, steel cylinder chambers, or cross section, are employed in all wet-process kilns and are now being used in dry-process kilns also. When exposed to the hot gases they rapidly reach a high temperature and, once they come into contact with the feed bed, transfer heat to the material by conduction..

A similar action takes place when heat transfers from the kiln wall to the feed. The flame radiates heat to the coating adhering to the kiln. Part of the heat then radiates to the bed, and part is transferred to the feed by conduction when the wall turns into the bed. Fig. 7.1 shows that the wall temperature is lowest when it emerges from the bed and highest immediately before it comes into contact with the feed bed. The slower the kiln speed and the smaller the cross-sectional loading (feed) of the kiln, the larger will be the temperature difference between these two points.

Other things being the same, there is an important relationship between the wall effect described above and the rotational speed of the kiln, as higher kiln speeds are more favorable than slow speeds for the heat exchange because of the smaller temperature difference between kiln wall and feed bed. These details must be taken into consideration by the kiln operator in deciding whether the kiln speed can be increased at any time.

In the preheater vessels this material-to-gas temperature relationship works differently. Here the heat exchange takes place more readily and thus the temperature difference between these two is much smaller. When discussing heat transfer in a kiln attention must be focused upon the decarbonation rate of the kiln feed for this requires by far the largest amount of heat in the entire process. In dry- and wet-process kilns, the total amount of heat for decarbonation must be transferred to the feed within the rotary

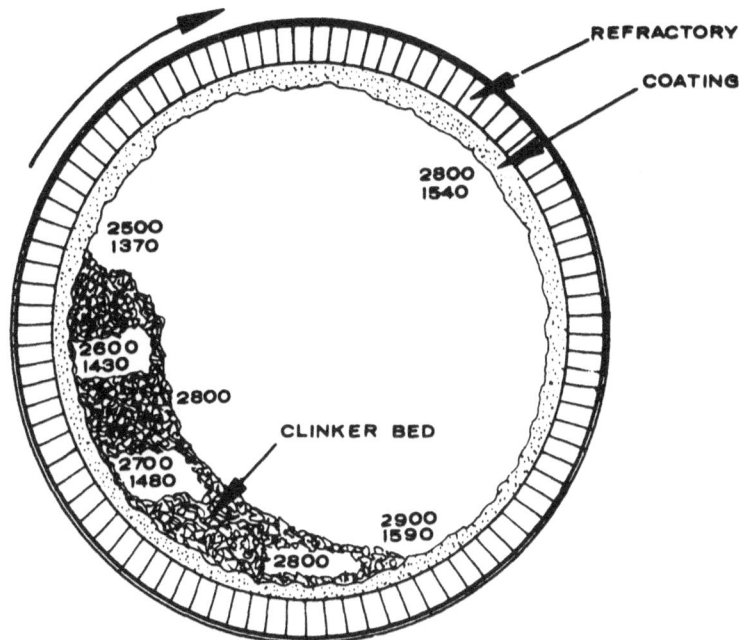

Fig. 7.1 Wall temperature of the kiln in the burning zone, reaching a maximum just before it comes in contact with the feed bed as the kiln turns, drops about 400 F (200 C) as it gives up much of its heat to the bed of material.

kiln itself. Since these cylinders are so inefficient in accomplishing this heat transfer, these types of kilns must be extremely long and wide. In a preheater kiln, the feed is calcined up to 30–35% in the preheater tower and thus the heat required, in terms of kcal/kg clinker, within the rotary kiln proper, is much less. This makes it possible to obtain either higher production rates on kilns of equal size or to maintain the same production rate by sizing the kilns much smaller. With the advent of the precalciner kiln this heat transfer efficiency has been taken a step further. On these kilns, the kiln feed is up to 90% calcined when it enters the kiln, in other words, most of the thermal work is done in the preheater tower. Theoretically, it would be possible to achieve even higher calcination rates in the precalciner but this is practically not possible as this would lead to the formation of viscous clinker phases and subsequent plug-ups in the lower stage of the preheater. Most operators therefore limit their precalciner to a maximum of 90% calcination prior to the feed entering the rotary kiln to prevent the material from sticking to the walls of the preheater tower.

Preheaters and precalciners have the advantage that the heat transfer is different from the transfer conditions found within the rotary kiln itself. In the kiln, the material bed runs practically undisturbed along the bottom of the kiln where the hot gases primarily heat only the surface layer of the bed. Hence, the contact between gas and material, i.e., the heat exchange, is relatively inefficient and takes a long time to complete. In the preheater and precalciner, the material is in suspension resulting in an immediate contact of feed particles with the hot gases.

This can be explained in simple terms when one looks at the time factor that is associated with calcination in these various kilns. A dry-process kiln requires approximately 45 m (150 ft) to achieve 90% calcination and the feed takes more than an hour to travel through this section of the rotary kiln. On the other hand, in the precalciner kiln, the same amount of thermal work is done in less than a minute. The feed in a precalciner and preheater kiln is much more uniformly calcined leading to better and more uniform kiln operating conditions than in the rotary kiln. Despite the sophisticated appearance of the preheater and precalciner kilns compared to the "old" wet and dry workhorses, they are easier to control from an operator's viewpoint. It can be safely stated that a good wet- or dry-kiln operator will also make a good preheater operator provided that he is properly retrained.

There have been a vast number of technical papers published by various authors on the subject of temperatures within different kiln systems. From these, comparative heat profiles can be drawn that approximate the temperatures typically found in such kilns. These are shown in Figs. 7.2 through 7.6.

A glance at these heat profiles shows the great difference in coating in relation to the overall kiln length. In the precalciner kiln, coating takes up 45–60% of the total kiln length whereas in the wet-process kiln this coating length is only 12–20% of the kiln length. Also significant is the temperature differential between material and gas in the preheat zones of a wet and a preheater kiln. It clearly shows the aforementioned superior heat exchange in a preheater cyclone when compared to the rotary cylinder.

Detailed discussions of the mathematics involved in the different types of heat transfer are beyond the scope of this book. It would not serve a kiln operator in the daily performance of the job. There is, however, one area of paramount importance to the kiln manager that deals with an essential part of kiln control; it is the heat transfer that takes place through the

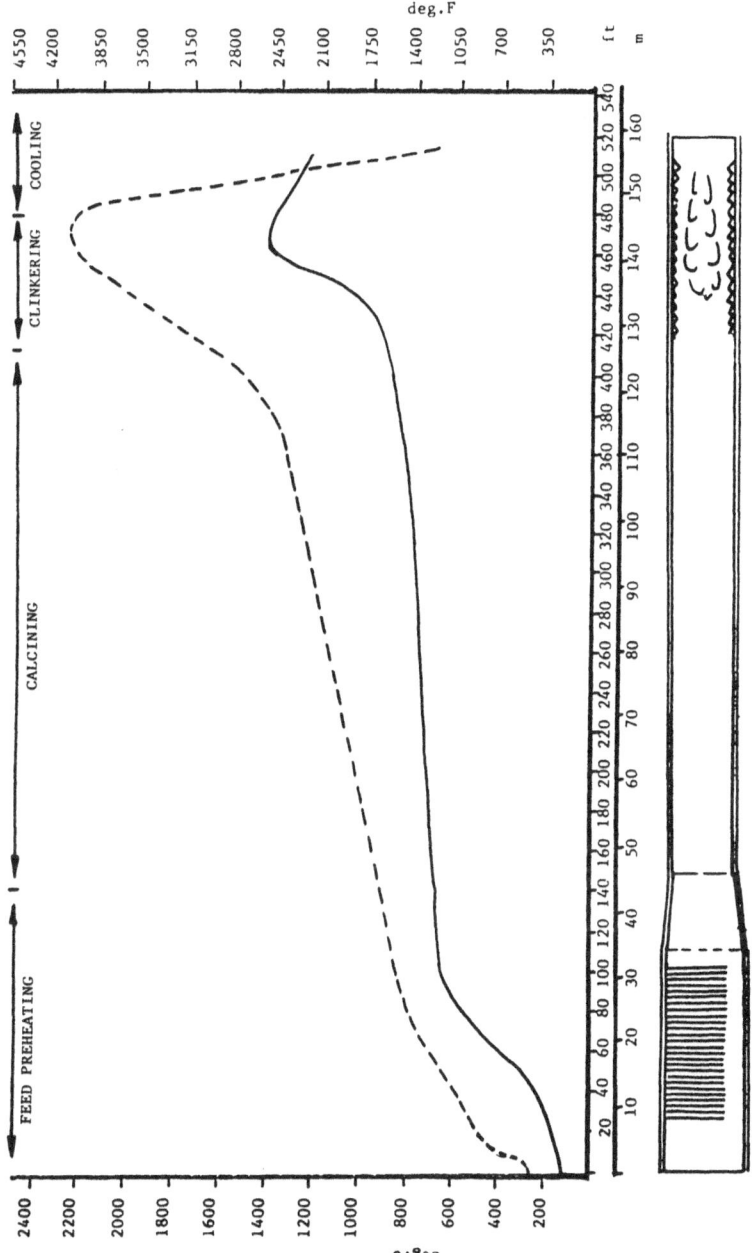

Fig. 7.2 Dry-process kiln material and gas temperatures.

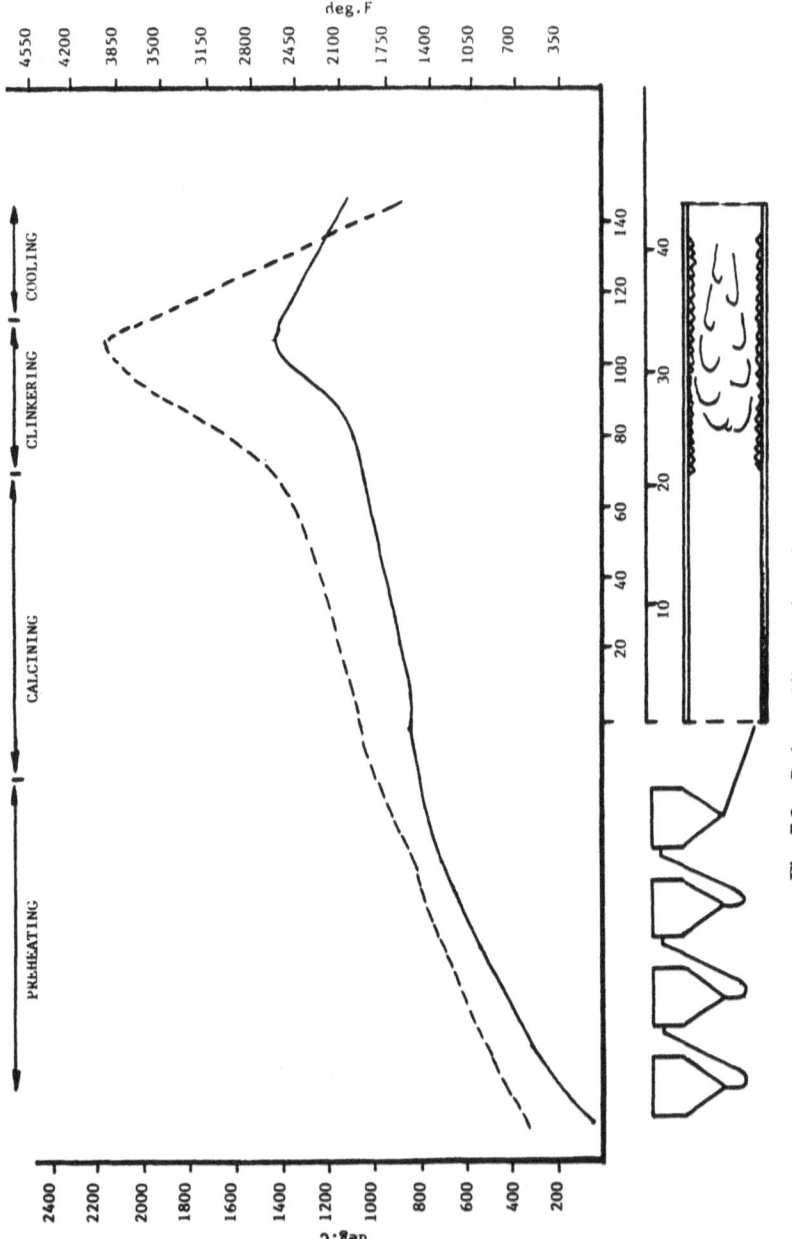

Fig. 7.3 Preheater kiln material and gas temperatures.

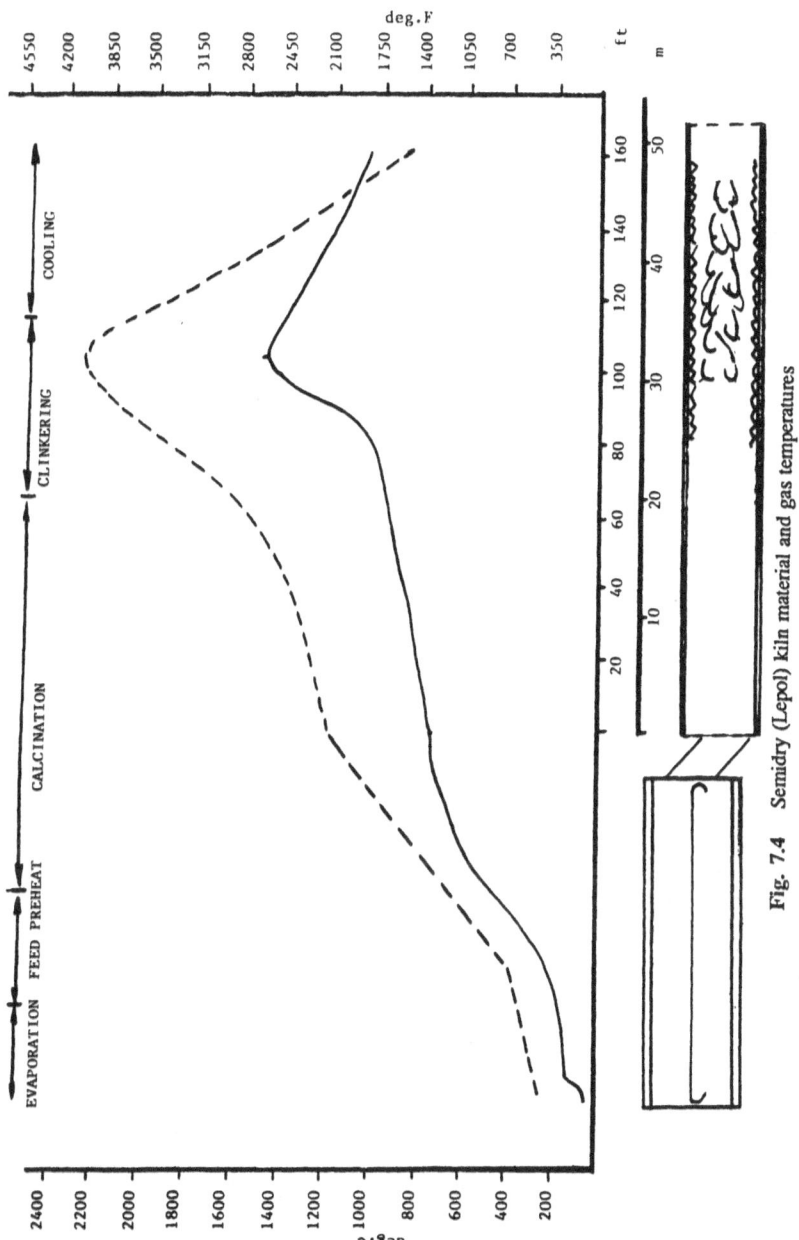

Fig. 7.4 Semidry (Lepol) kiln material and gas temperatures

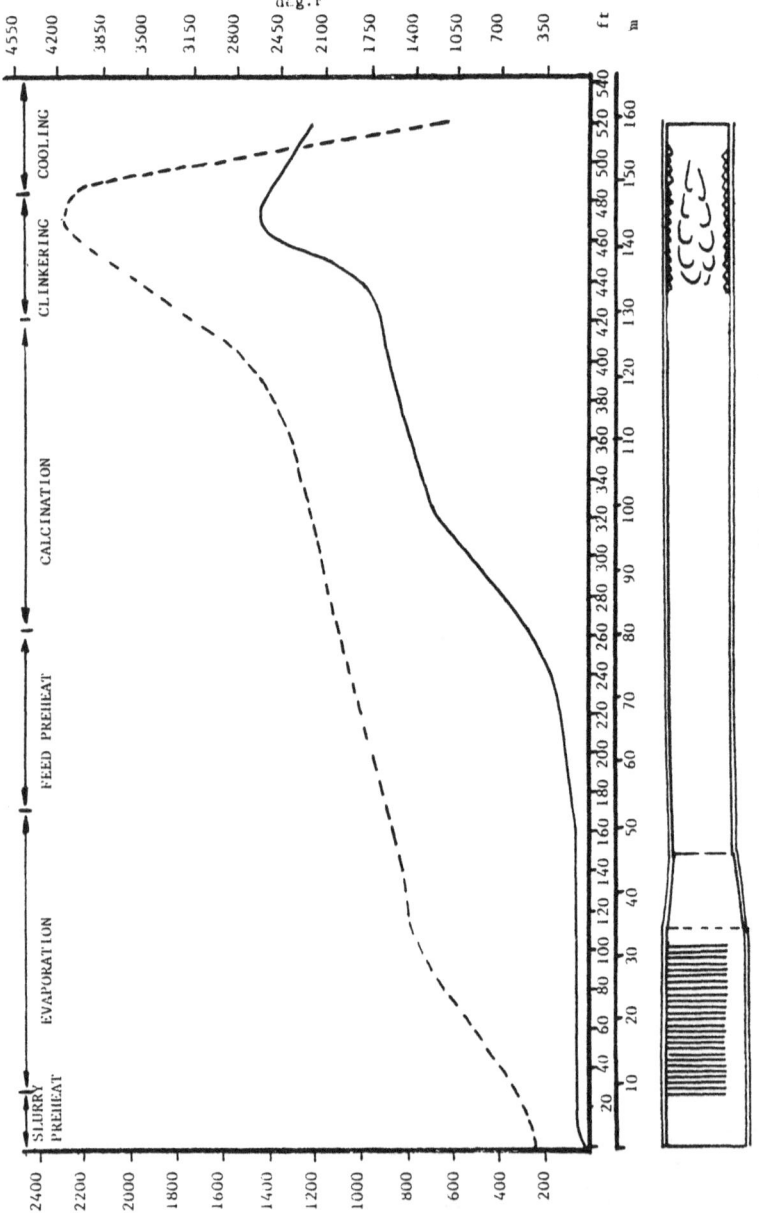

Fig. 7.5 Wet-process kiln material and gas temperatures.

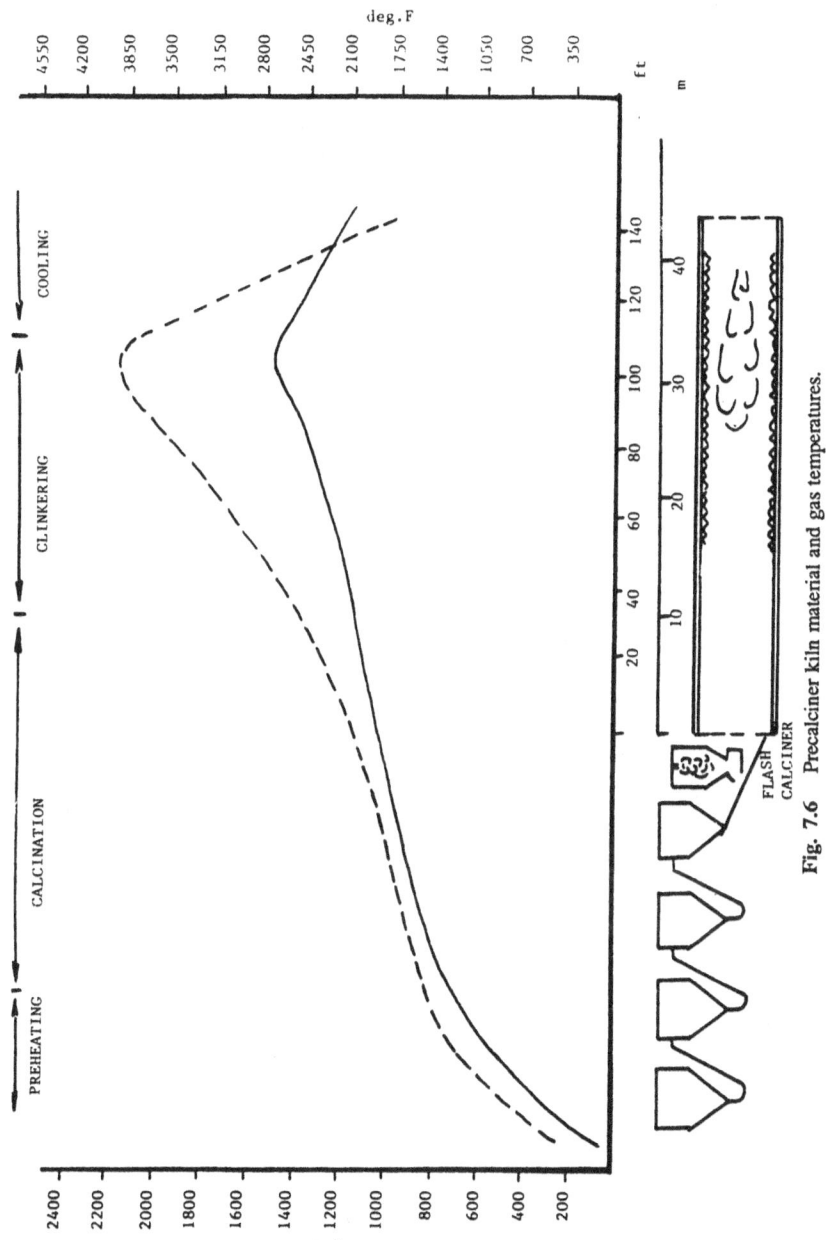

Fig. 7.6 Precalciner kiln material and gas temperatures.

refractory walls of a rotary kiln which in turn governs the temperature of the kiln shell. Kiln managers frequently have to deal with questions that involve the selection of appropriate refractory types and sizes to minimize heat losses through the kiln wall and to hold the kiln shell temperature within acceptable limits.

Some of the fundamental steps and formulas that allow for the determination of shell temperature and heat losses when different refractories are used in a rotary kiln are discussed below.

7.4   FUNDAMENTAL FORMULAS FOR HEAT-TRANSFER CALCULATIONS

The fundamental law for heat transfer

$$Q = K_1 A \; \frac{t_1 - t_2}{x} \tag{7-1}$$

only applies to conditions where K_1 and A are independent of t and x. However, as with most other materials, the heat-transfer coefficient K_1 for refractories changes with the temperature. For this reason, the value K_1 must always be accompanied with a statement as to the temperature at which the value is expressed. For practical purposes in refractory heat-transfer calculations, the relationship K_1 vs t in most instances is linear and therefore the heat-transfer equation can be written in the form:

$$Q = KA \; \frac{t_1 - t_2}{x} \tag{7-2}$$

where:

Q = quantity of heat flow per unit time
K = average of heat-transfer coefficients K_1 over the temperature range $t_1 - t_2$
A = heat-transfer surface area
t_1 = hot-face temperature of refractory wall
t_2 = cold-face temperature of refractory wall
x = refractory thickness

Either English, Metric, or S.I. units can be applied in this formula. The heat-transfer coefficient is commonly expressed in the following terms:

English system of units:

$$K = \cfrac{\dfrac{\text{Btu}}{\text{h}}}{\text{ft}^2 \cdot \dfrac{°\text{F}}{\text{in.}}} = \text{Btu-in./ft}^2\text{h}°\text{F}$$

Metric system of units:

$$K = \cfrac{\dfrac{\text{kcal}}{\text{h}}}{\text{m}^2 \cdot \dfrac{°\text{C}}{\text{m}}} = \text{kcal/mh}°\text{C} \qquad (7\text{-}3)$$

(Note: In the metric system, K is often denoted as λ)

International system of units:

$$K = \cfrac{\dfrac{\text{kJ}}{\text{h}}}{\text{m}^2 \cdot \dfrac{°\text{C}}{\text{m}}} = \text{kJ/mh}°\text{C}$$

To simplify heat-transfer calculations, it is customary to use the thermal resistance value R of the refractory wall, since R is a function of both K and x.

$$R = \frac{x}{k} \qquad (7\text{-}4)$$

In the above formula, K and x have the same meaning as in Eq. (7-2). Using the R value instead of K has the advantage for the engineer to rapidly develop graphical solutions to problems in finding the heat loss, Q and the cold-face temperature, t_2 when t_1, K, x, and the ambient air temperature are known.

It must, however, be stressed that heat-transfer calculations for rotary-kiln linings only deliver approximate results since several assumptions and estimates must enter into the calculations. This can be explained in more detail by viewing the following heat-transfer diagram of a typical cement kiln wall.

If the direction of the heat flow is followed, problems will immediately be experienced since it is difficult to accurately state the hot-face temperature of the lining (t_1) because both the interior temperature, t_k and the coating thickness are unknown entities in practial application. This is especially true for the burning zone but applies as well to all other zones of the kiln. Furthermore, although refractory manufacturers supply heat-transfer coefficients (K) on new bricks, it is known that the thermal conductivity properties can undergo changes due to the interaction of kiln gases and the kiln-feed components with the refractory hot face during its normal service life in the kiln.

Step 1: The first calculations are made by estimating the brick hot-face temperature and stating the desired shell temperature one is striving for. From these two, the estimated temperature gradient ($t_1 - t_2$) is calculated. This is a preliminary step to find the appropriate thermal-conductivity factor, K.

Step 2: In order not to introduce any more errors into the calculations, it is customary to use the logarithmic mean temperature of the brick for calculations involving linings in cylindrical vessels.

$$\text{log mean temperature} = 0.434 \; \frac{t_1 - t_2}{\log \dfrac{t_1}{t_2}} \qquad (7\text{-}5)$$

When detailed information for this heat gradient are not available, Table 7.1 can be used as a general guideline.

Step 3: From the manufacturer's data sheet the thermal-conductivity factor (K) that corresponds to the above calculated log mean temperature is found (or refer to Fig. 7.2).

Step 4: The wall resistance factor, R, is calculated:

$$R = \frac{x}{K} \qquad (7\text{-}6)$$

where
x = lining thickness.

Step 5: Read the desired results for t_2 and Q directly from Fig. 7.7 or Fig. 7.8 depending on the R factor calculated above.

The amount of heat lost by radiation and convection from the shell can vary considerably from kiln to kiln even if the shell temperature is the same. The values obtained by using the attached graphs are applicable to conditions where the surrounding air temperature is taken as a constant 21 C (70 F) in still air. It is also known that different colors and shell surface characteristics can produce different emissivities for radiation from the shell. For example, a kiln shell painted with aluminum paint shows an emissivity of ≈ 0.60 whereas a typical steel shell, slightly oxidized, shows an emissivity of between 0.80–0.88. For the purpose of this discussion, a value of 0.82, is used as a constant.

Convection losses, likewise, can also vary depending on the characteristics of the fluid film immediately adjacent to the kiln shell. Shell cooling fans, wind velocities, and temperatures of the surrounding ambient air play an important role in the amount of heat lost by convection from the shell. A roof over a kiln section, another operating kiln nearby, and a variety of other reasons can cause modifications in the amount of heat lost from a given kiln section. All these intangible factors reemphasize that results obtained by this method of calculation are to be viewed as approximate values only.

The graphs in Figs. 7.7 and 7.8 have been developed by using two fundamental heat-transfer equations. First, for the heat transferred through the wall by conduction:

$$Q_1 = \frac{t_1 - t_2}{R} \tag{7-7}$$

Second, for the heat lost from the kiln shell to the surrounding air by radiation and convection:

$$Q_2 = 0.472 \, E \left(\frac{T_{abs}}{100}\right)^4 + 0.5254(t_1 - t_2)^{\frac{5}{4}} \tag{7-8}$$

Finally, in drawing the heat-transfer curves on the graph, the individual reference points must satisfy the relationship expressed in Eqs. (7-7) and (7-8). In other words:

$$Q = Q_1 = Q_2$$

where:

Q_1 =   amount of heat transferred by conduction through the wall $(kcal/m^2 \cdot h)$

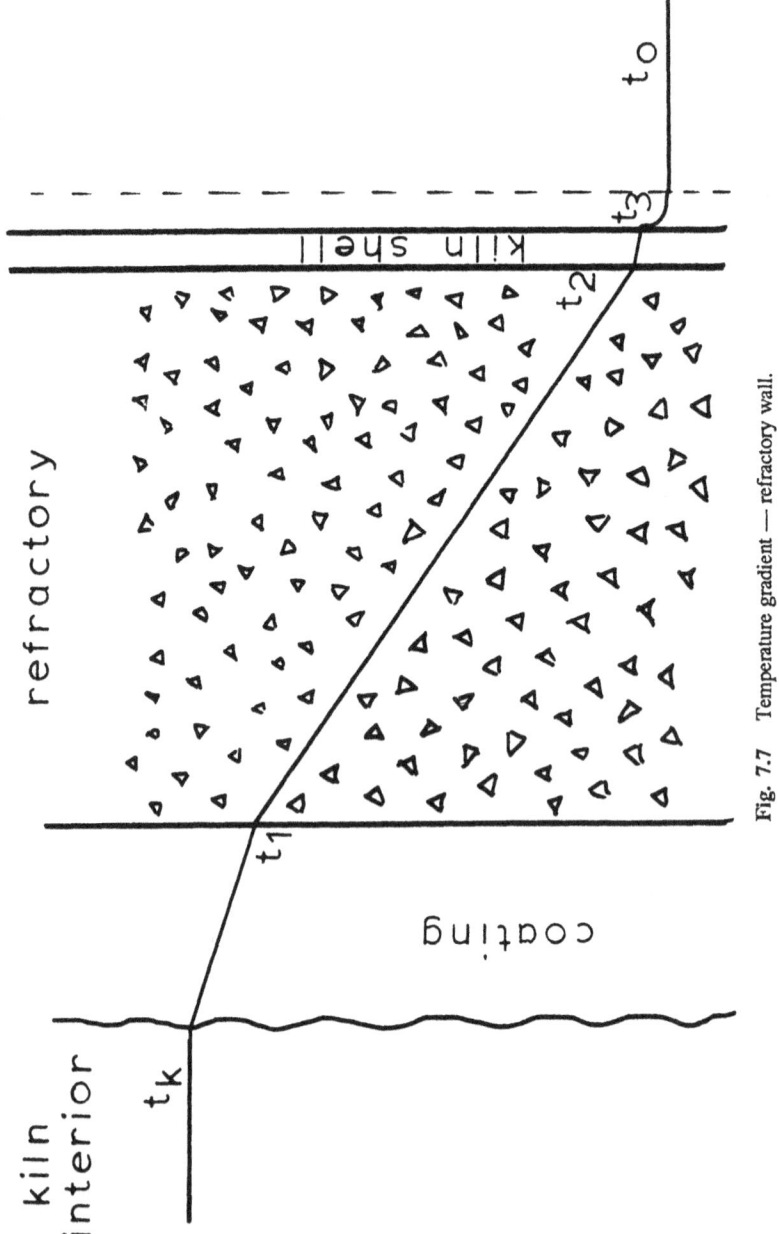

Fig. 7.7   Temperature gradient — refractory wall.

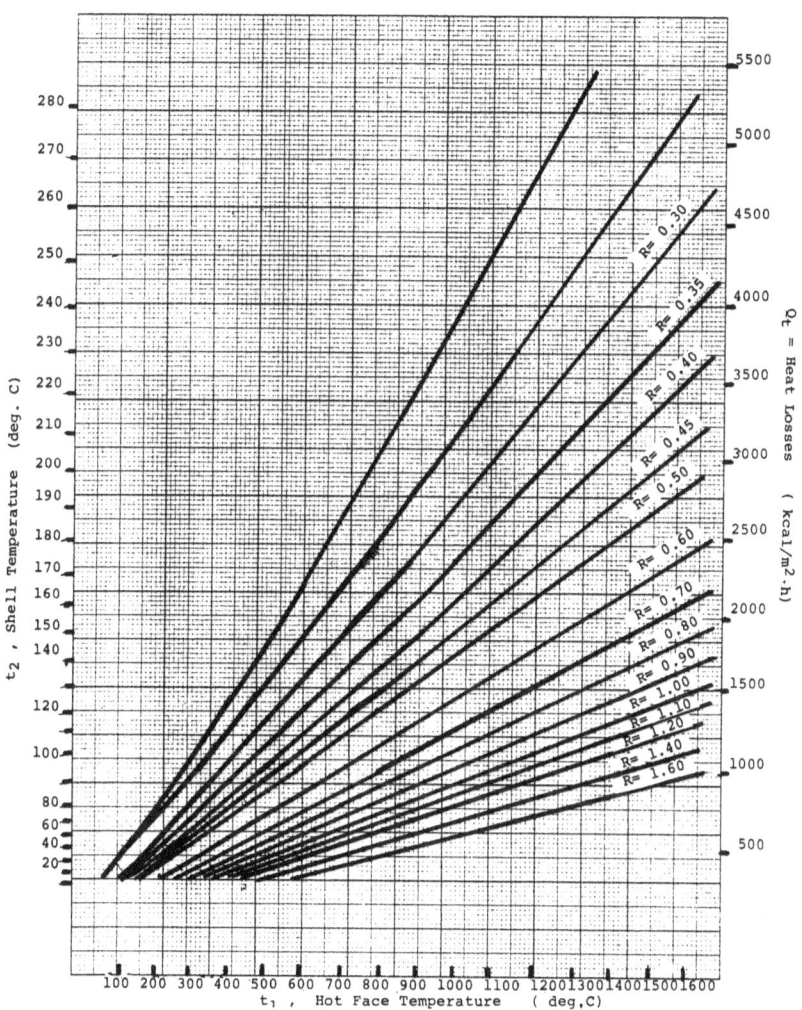

Fig. 7.8   Shell temperature and heat losses at $R = 0.20$ to 1.60.

Q_2 =   radiation and convection losses from the kiln shell
              ($kcal/m^2 \cdot h$)
t_1 =   hot-face temperature (°C)
t_2 =   shell temperature (°C)
R =   thermal resistance of wall
E =   emissivity = constant: 0.82
T_{abs} =   absolute temperature.

Example 1: A rotary kiln is being equipped with a 225-mm high, semi-insulating lightweight lining in the upper part of the calcining zone. What is the approximate shell temperature and heat loss?

TABLE 7.2

TYPICAL HOT-FACE TEMPERATURES AND TEMPERATURE GRADIENTS IN CEMENT KILN WALLS (estimates)

	Hot-face temp. t_1	Shell temp. t_2	Δt ($t_1 - t_2$)	Log mean temp.
Kiln Hood				
Wet-process kiln	650	160	490	350
Dry-process kiln	875	200	675	460
Suspension P.h.	1100	230	870	560
Precalciner kiln	1200	245	955	600
Grate Coolers				
Lower part	300	95	205	180
Upper part	1000	210	790	510
Kiln Outlet Zone				
Without coating	1300	410	890	770
Light coating	1050	365	685	650
Medium coating	650	280	370	440
Heavy coating	450	225	225	330
Burning Zone				
Without coating	1475	440	1035	860
Light coating	1100	360	740	660
Medium coating	750	280	470	480
Heavy coating	500	240	260	350

TABLE 7.2 *(cont'd.)*

TYPICAL HOT-FACE TEMPERATURES AND TEMPERATURE GRADIENTS IN CEMENT KILN WALLS (estimates)

Transition Zone				
Without coating	1300	410	890	770
Light coating	1050	365	685	650
Medium coating	650	280	370	440
Heavy coating	450	225	225	320
Calcining Zone				
Lower part:				
with insulation	1230	160	1070	530
without insulation	1230	240	990	610
Middle part:				
with insulation	1100	150	950	480
without insulation	1100	220	880	550
Upper part:				
with insulation	1000	140	860	440
without insulation	1000	210	790	510
Preheat Zone				
Suspension P.h.:				
with insulation	925	140	785	420
without insulation	925	200	725	470
Dry kiln:				
with insulation	875	130	745	390
without insulation	875	190	685	450
Wet kiln:				
with insulation	810	125	685	370
without insulation	810	185	625	420
Chain Zone				
Wet-process kiln	300	95	205	180
Dry-process kiln	575	140	435	310
Feed Inlet Zone				
Wet-process kiln	205	80	125	130
Dry-process kiln	425	130	295	450
Suspension P.h.	875	205	670	460

Solution:

a) From Table 7.2, $t_1 = 1000$; t_2 estimated $= 140$ and log mean temperature of the temperature gradient is 440 C.

b) From Fig. 7.8: K at 440 C = 0.36.

c) $R = 0.225/0.36 = 0.625$

From Fig. 7.9: Starting vertically from hot-face temperature 1000 C, find the location where this straight vertical line intercepts the heat transfer curve near $R = 0.60$. Horizontal to the left, the shell temperature is read as 120° and horizontal to the right of this intercept the heat loss is read as 1450 kcal/m² · h.

Example 2: What would the heat losses and shell temperatures be if in Example 1 a 40% fireclay brick of the same thickness is used?

Solution:

a)　　$t_1 = 1000$, $t_2 = 210$, log mean temperature, 510 C

b)　　K at 510 C = 0.94

c)　　$R = 0.225/0.97 = 0.24$

d)　　From Fig. 7.3, vertical from 1000 C and horizontal from intercept near $R = 0.25$, a shell temperature of 213° and a heat loss of 3270 kcal/m² · h are obtained.

Example 3: An engineer desires to use a basic magnesite-chrome brick over the second tire section which is located at the transition zone. What height of lining is required if the shell shall not exceed 370 C? It is projected that this area will contain a light coating.

Solution:

a) From Table 7.2, $t_1 \approx 1050$ C.

b) From Fig. 7.9, finding intercept t_1 1050 and t_2 370: $R = 0.075$

c) Log mean temperature = $0.434 \dfrac{1050 - 320}{\log \dfrac{1050}{320}}$ = 613 C

d) From Fig. 7.2, K for 80% MgO brick at log mean temperature of 613 C = 2.7

e) From Eq. (7-4), $R = x/K$, determine x:

$x = 2.7 \times 0.075 = 0.203 = 200$ mm

Thus, to maintain the temperature below 370 C, the lining thickness should be 200 mm.

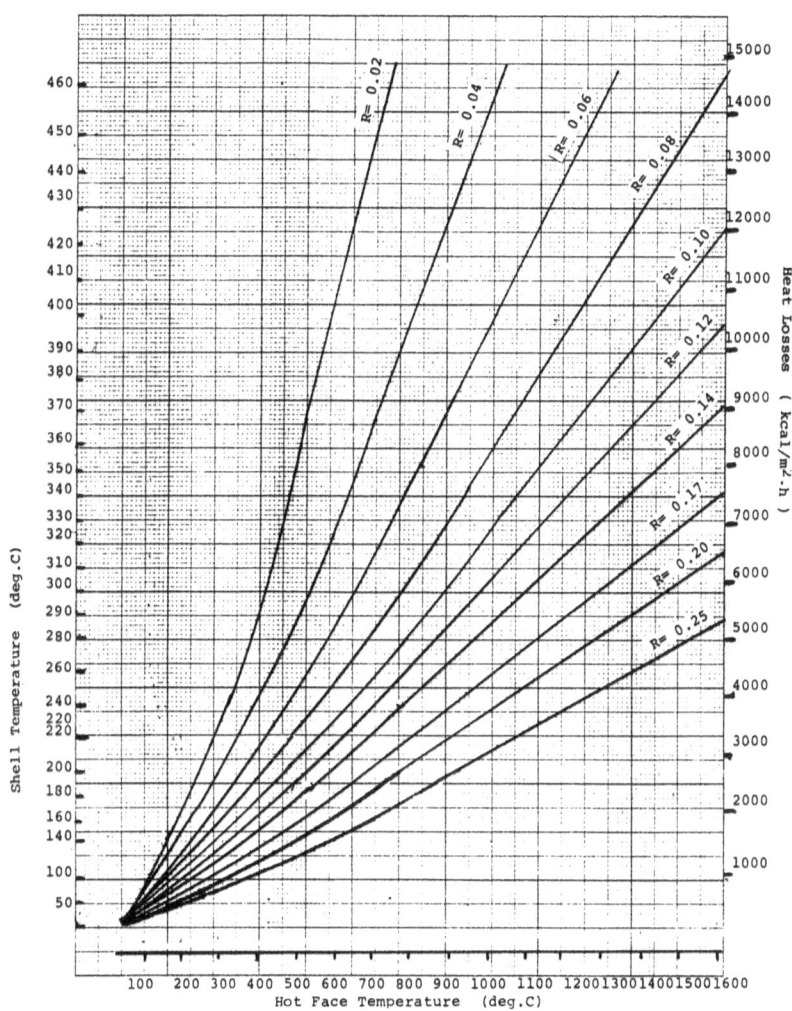

Fig. 7.9   Shell temperature and heat losses at $R = 0.02$ to 0.25.

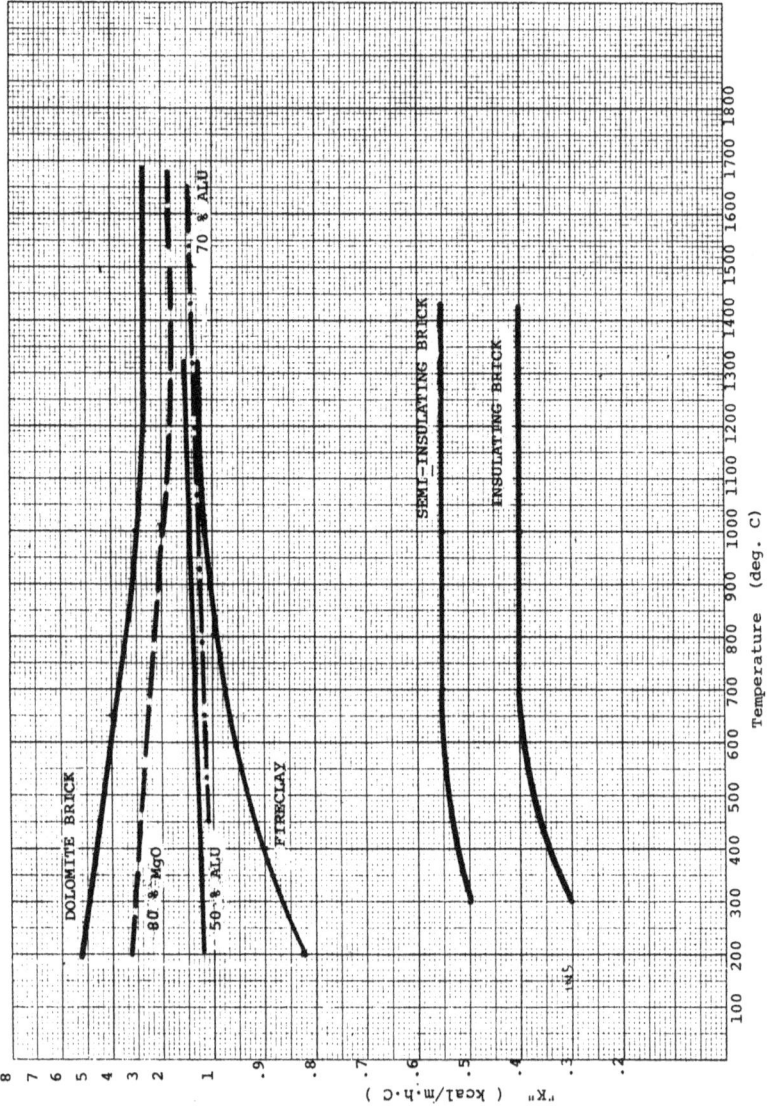

Fig. 7.10   Coefficients of thermal conductivity (kcal/m . h . °C).

Example 4: A kiln shell usually shows a faint dark red color at night when its temperature registers approximately 460 C.   Referring to Example 2, at which brick thickness can such a red spot be expected with 40% alumina fireclay brick?
Solution:

a) From Table 7.2, t_1 = 1000

b) From Fig. 7.9, finding intercept t_1 1000 and t_2 460: R = 0.038

c) Log mean temperature = $0.434 \dfrac{1000 - 460}{\log \dfrac{1000}{460}}$ = 694 C

d) From Fig. 7.7: K for fireclay brick at log mean temperature of 694 C = 0.99

e) From Eq. (7-4): $R = x/K$

x = 0.99 X 0.038 = 0.038 = 38 mm

Thus, the shell could show a faint dark red color when the brick thickness has worn down to approximately 40 mm.

References

1. De Keyser, W. L., 1955.   Reactivity in the solid state between the oxydes of the cement system.
2. Margin, G.   1932.   Chemical eng. and thermodynamics applied to the cement rotary kiln.

8.

Heat Balances

Before modern developments in kiln technology, there were efficient kiln operators that never had a complete understanding of what actually took place within a given kiln system. These operators were proficient in controlling the temperatures, pressures, and material flow rates and did their job so well that some of them remained kiln burners for the rest of their career. Still today, there is no absolute requirement for an operator to be knowledgeable about heat balances. These are usually of more interest to the plant engineer.

Nevertheless, it is appropriate to devote a separate chapter to this topic in order to gain a deeper understanding of the process being controlled. The operator who can keep calm and adjust to unusual situations without losing composure deserves credit. For most erratic actions there is a plausible explanation and these explanations can become legitimate reasons when a deeper understanding of what really goes on inside a kiln is reached.

Kiln operating efficiency covers a broad spectrum of considerations, each one having equal importance. On one occasion interest in the output efficiency may be of concern, at another time concern might shift to cooling efficiency because the clinker cooler might not be doing the kind of work it was designed to do. On an almost daily basis consideration must be given to how efficiently the fuel, given to the kiln to do the true thermal work within the system, is being used. This chapter focuses on the fuel factor because it is one of the major cost factors in the manufacturing process of cement. It might surprise some kiln operators when they learn that the fuel bill for the kiln alone can amount to 3–4 million dollars in a year. Not so long ago, fuel prices were so cheap in the United States that nobody was concerned with the quantity of fuel burning in the kiln. Today, this aspect can not be ignored.

A heat balance, simply stated, consists of compiling all the heat that is given to the kiln and then comparing this total to the total of thermal work done and heat losses that occur in the system. Whatever heat put into the kiln (INPUT) must be accounted for in one way or another by the heat that goes out of the system (OUTPUT). To do this requires actual testing of the system under normal operating conditions. Some plants have done this by means of very elaborate and sophisticated instruments, others have used average operating data from the kiln operator's log to compile and calculate heat balances.

The following shows three different heat balance models, one each for a wet, a dry, and a preheater kiln. In comparing these the differences in overall specific heat consumption, heat requirements, and heat losses between these kilns can be observed.

For the uninitiated, it should be noted that the actual heat required for clinkering the feed, i.e., the true thermal work required, is only about 431 kcal/kg (1.55 MBtu/sh.ton). In the kiln models presented, the wet-process kiln with its 5.022 MBtu/sh.ton (1396 kcal/kg) specific heat consumption is thus only about 31% efficient, the dry-process kiln at 4.295 MBtu/-sh.ton (1194 kcal/kg) is slightly better at 37%. The best fuel efficiency at 50% efficiency is found in the preheater kiln with 3.106 MBtu/sh.ton (863 kcal/kg).

The best reported efficiencies are near the 2.7 MBtu/sh.ton (750 kcal/kg) levels all of which are found in either suspension preheater or precalciner kilns. This would represent a 57% efficiency which by itself approaches close to the optimum possible limit since the most efficient heat-transfer engine in the world works with an efficiency of around 68%. In short, there are limits to how efficient a kiln system can become. When kilns approach these limits it becomes increasingly more difficult and almost impossible to squeeze an extra kJ, kcal, or Btu out of the system to perform useful thermal work. Expressed in another manner, it most probably would be easier to lower the fuel consumption on a wet-process kiln from, e.g., 6.8 MBtu/t to 5.9 MBtu/t than to squeeze another 50,000 Btu/t out of it when the kiln is already operating at 4.95 Btu.

It should be of interest to the reader to compare the right column, which expresses the heat output in percentages, for each of these three kiln models. Each one of these given percentages are revealing by themselves. But there is more to it than just testing the kiln to answer the question:

"Where does all that heat go?" Heat balances are the foundation for an engineer to identify the opportunities for possible heat savings. There are several areas where heat could possibly be saved to make a kiln operate more efficiently in matters of energy consumption. Some of these potentials are:

1. **Lower kiln exit gas losses by:**
    - installing more chains and inserts such as trefoils, lifters, and slurry preheaters on wet- and dry-process kilns.
    - reducing the clinkering zone and extending the length of the calcining zone.
    - operating the kiln at optimum oxygen levels with minimum excess air in the kiln exit gases.
    - designing proper fuel burners that deliver optimum flame shapes and temperatures.
    - on dry kilns: using the high exit gases to produce steam and generate electrical power or using this excess heat for drying raw materials.
    - on preheater kilns: adding additional preheater cyclones or improving the heat exchange between gas and solids within existing cyclones.
    - operating preheaters with only as much bypass as is absolutely necessary to control the alkali in the clinker.
    - equipping the dry-process kiln with a single-stage preheater.

2. **Lower the heat losses associated with evaporating the free water from the kiln feed in wet-process kilns by:**
    - using slurry thinners.
    - grinding the slurry consistently to the lowest possible moisture content.
    - using filter presses to dewater the slurry before it enters the kiln.

3. **Lower the dust in the exit gases by:**
    - improving the chain system and other insert designs in dry- and wet-process kilns to prevent escape of dust at the feed end.
    - optimizing kiln operating procedures and kiln loading to minimize turbulence and excessive gas velocities within the kiln itself.

TABLE 8.1

HEAT BALANCE FOR THE WET-PROCESS KILN

Heat input	1000's Btu/t	Percent	Heat output	1000's Btu/t	Percent
Combustion of fuel	4845.84	96.5	Theoretical heat required	1534.16	30.5
Sensible heat in fuel	4.25	0.1	Exit gas losses	646.45	12.9
Organic matter in feed	—	—	Evaporation of moisture	1925.64	38.3
Sensible heat in feed	97.86	1.9	Dust in exit gas	9.73	0.2
Sensible heat in cooler air	65.17	1.3	Clinker discharge	48.68	1.0
Sensible heat in primary air	7.99	0.1	Cooler stack losses	163.29	3.3
Sensible heat in infiltrated air	0.00	0.1	Kiln shell losses	582.71	11.6
			Losses due to calcination of wasted dust	35.03	0.7
			Unaccounted losses	76.44	1.5
Total	5022.11	100.0	Total	5022.11	100.0

Note: Unaccounted losses are calculated by difference to make the two sides equal.

TABLE 8.2

HEAT BALANCE FOR THE DRY-PROCESS KILN

Heat input	Btu/t	Percent	Heat output	Btu/t	Percent
Combustion of fuel	4144.12	96.5	Theoretical heat required	1573.19	36.6
Sensible heat in fuel	3.63	0.1	Exit gas losses	1189.61	27.7
Organic matter in feed	—	—	Evaporation of moisture	258.10	6.0
Sensible heat in feed	68.43	1.6	Dust in exit gas	11.14	0.3
Sensible heat in cooler air	70.30	1.6	Clinker discharge	52.62	1.2
Sensible heat in primary air	7.38	0.2	Cooler stack losses	508.24	11.8
Sensible heat in infiltrated air	0.94	—	Kiln shell losses	521.17	12.1
			Losses due to calcination of wasted dust	15.85	0.4
			Unaccounted losses	164.87	3.8
Total	4294.80	100	Total	4294.80	100

Note: Unaccounted losses are calculated by difference to make the two sides equal.

TABLE 8.3

HEAT BALANCE FOR THE SUSPENSION PREHEATER KILN

Heat input	Btu/t	Percent	Heat output	Btu/t	Percent
Combustion of fuel	2979.00	95.9	Theoretical heat required	1541.500	49.6
Sensible heat in fuel	2.61	0.1	Exit gas losses	427.370	13.8
Organic matter in feed	—	—	Evaporation of moisture	202.210	6.5
Sensible heat in feed	52.11	1.7	Dust in exit gas	1.110	—
Sensible heat in cooler air	66.46	2.1	Clinker discharge	56.610	1.8
Sensible heat in primary air	5.26	0.2	Cooler stack losses	527.800	18.4
Sensible heat in infiltrated air	0.51	—	Kiln shell losses	150.722	4.9
			Losses due to calcination of wasted dust	5.321	0.2
			Unaccounted losses	148.302	4.8
Total	3105.94	100	Total	3105.94	100

Note: Unaccounted losses are calculated by difference to make the two sides equal.

- returning (reclaiming) the maximum possible amount of kiln dust to the kiln by means of insufflation, leaching, or other means.
- enlarging the feed end and the shell in the chain section to lower the gas velocity in these areas.

4. **Lower the clinker-discharge temperature by:**
  - operating the clinker cooler at the minimum necessary air-flow rates but doing the maximum possible cooling work. (This is primarily governed by the design of the cooler and its components which, even today, leaves a lot to be desired and offers plenty of room for improvement.)

5. **Lower the cooler stack (exit) losses by:**
  - (same as mentioned above)
  - recycling excess cooler air from the lower cooler area to the upper cooler compartment fans. (Watch the wear on the fan blades.)
  - reclaiming excess cooler air for drying of raw materials, coal grinding, heating of fuel oil or for combustion air in the auxiliary flash furnace of a precalciner.
    Note: Cooler stack losses are closely related to amount of combustion air needed in the kiln. Unfortunately, the more efficient (in specific fuel consumption) a kiln becomes, the higher the cooler exit losses will be. The heat-balance examples show this clearly where the inefficient wet-process kiln shows 3.3% cooler losses compared to the more efficient preheater kiln which shows 18.4% losses.

6. **Lower kiln radiation losses by:**
  - using insulating refractories in the areas above the burning zone. (Watch the maximum permissible service temperature of these bricks.)
  - selecting the proper mix design and refractory that delivers heavy coating in the burning zone without leading to excessively large ring formations.
  - using back-up insulation on all equipment linings outside the rotary kiln, e.g., the air ducts, kiln hoods, cooler walls, and the preheater tower.

7. **Lower cold-air inleakage into the kiln system by:**
  - closing up all unnecessary openings in the kiln hood, cooler, and kiln feed end.

- providing for efficient kiln seals both at the lower as well as the upper end of the kiln.
- operating the kiln with as high a primary air temperature as is safe and possible.

Most of the above-mentioned improvement factors are part of the inherent designs of the kiln systems where a kiln operator has usually little influence. It is primarily a concern to the engineering department. There are, however, many other factors, directly under the control of the operator, that will have a bearing on the overall kiln-energy efficiency. These factors are discussed in more details with the discussion of the actual control functions of the kiln.

When striving to lower the energy requirements on a kiln, it must be remembered that energy saved in one particular area might lead to a higher energy loss in another. For example, adding more chains on a dry or wet kiln might help to reduce the kiln exit-gas losses but invariably raises the kiln drive-power requirements. An engineer must therefore always balance the obtained savings by the resultant, possibly higher energy requirements this action might cause before proceeding with the project. From an operator's viewpoint it might be more meaningful to explain the theoretical savings obtainable in monetary terms.

Using a computer program, a simulation was run on a wet-process kiln to show the effect on overall fuel efficiency when certain changes in the process variables were made. Here then are the results of this particular run:

Wet Kiln (129.5 × 3.43 m)

| Annual clinker production: | 212,800 metric tons |
| Fuel costs: | 0.5418 $ per million kcal |

Variable	Change made	kcal/kg saved	Annual $ saved
Exit gas temperature	20 C lower	23.86	43,200
Moisture in slurry	2.5% lower	50.94	92,500
Clinker discharge temperature	40 C lower	6.96	12,600
Cooler stack temperature	20 C lower	12.95	23,500
Kiln-shell temperature	20 C lower	16.11	29,200
Total potential savings		110.82	201,000

This example shows the resultant savings for only one relatively small kiln. There have been reports where up to a million dollars annually have been saved on larger kilns after extensive modifications were made on existing systems. Naturally, in such plants energy conservation was a high-priority item and a well-defined conservation program must have been in place. The middle 1970's was a difficult period when the so-called energy crisis occurred. At the time of this writing (1985) fuel costs have gone down and there appears to be an energy glut on the world markets. Unfortunately, fuel conservation efforts seem to have taken a back seat again in many cement plants because of the apparent easy availability of fuels. This false sense of security should not exist for the simple reason that world market conditions change and when a kiln has a potential for fuel conservation it should be implemented regardless of the present availability of fuel.

9.

The Chemistry of Kiln Feed and Clinker

9.1 RAW MATERIALS

The basic ingredients for portland cement consist of limestone, sea shells, marl, or chalk, that provide the calcareous components; clay, shale, slate, or sand, to provide the silica and alumina; and iron ore, mill scale, or similar material to provide the iron components. The number of raw materials required at any one plant depends upon the composition of these materials and the types of cement being produced. To effect the proper blend, raw materials are continually sampled and analyzed, and the proportions adjusted as they are blended together.

After being excavated in the quarry or mine, limestone is first passed through the primary crusher then to the secondary crusher where it is reduced to about 3/8 in. in size. At this point other raw materials are blended with the limestone and the blend is conveyed to raw storage piles. Samples, mentioned above, are obtained at this point and immediately analyzed. In a modern plant this sampling and testing is the source of data fed into a digital computer controlling composition of stored and blended raw feed.

In the dry process, the material is now removed from the blending piles and delivered to the raw grinding mills, where it is reduced in size until about 90% passes the 200-mesh screen.

In the wet process, the raw feed is transferred from raw storage piles to the grinding mills, which are substantially the same as the ball, tube, or compartment mills used for dry grinding. Introduction of water into the mill along with the feed results in the formation of a slurry. After grinding, dry kiln feed or slurry is drawn from storage and fed into the rotary kiln.

115

9.2   CHEMICAL AND PHYSICAL PROPERTIES

Usually, any one constituent of the blended kiln feed can be found in more than one of the raw materials. For example, typical raw materials might contain key oxides in the proportions shown in Table 9.1.

TABLE 9.1

TYPICAL COMPOSITION OF RAW MATERIALS

	CaO	SiO_2	Al_2O_3	Fe_2O_3	MgO	*Loss*
Limestone (chalk)	52.0	5.7	0.8	0.3	0.4	40.4
High-silica limestone	33.6	36.8	1.8	0.6	0.5	26.4
Cement rock	40.0	18.0	5.0	1.5	2.0	32.0
Blast-furnace slag	35.5	33.1	9.1	0.9	16.4	2.1
Shale	3.2	53.8	18.9	7.7	2.2	13.1
Sand	0.8	70.0	15.0	5.0	0.2	8.6
Clay	0.5	61.0	16.9	12.4	0.4	7.8
Iron ore	—	6.7	1.4	89.7	0.4	0.2
Steel-mill scale	—	2.5	1.1	89.9	—	4.0

From such typical raw materials, a plant chemist tries to obtain a kiln-feed mix that contains a predetermined oxide amount of calcium (CaO), silica (SiO_2), alumina (Al_2O_3) and iron (Fe_2O_3).   In some locations, mixing of only two or three different raw materials accomplishes this, whereas in some other plants it might need up to four or five different materials to achieve the same results.

In addition to these basic oxides, raw materials also contain a certain percentage of so-called impurities which show up in the kiln-feed mix. *Magnesia* (MgO), in some plants can amount to levels of 4.2% in the mix which, if not properly controlled, can lead to unsound (expansion) cement. Magnesia acts as a flux at sintering temperatures which renders the burning slightly easier.   However, a magnesia-rich kiln feed tends to "ball" easily in the burning zone which, from an operator's viewpoint, is considered an undesirable property.

Clinker, made from magnesia-rich feed must be very rapidly cooled once it has been burned, to guard against production of unsound clinker. Plants faced with this problem usually locate the burning zone very close to the discharge end of the kiln and have quick quench compartments at the inlet to the cooler.

Oxides of *potassium* (K_2O) and *sodium* (Na_2O), commonly referred to as alkalies, are impurities that not only have a deleterious effect on cement quality but can pose considerable operating problems particularly in a preheater kiln. Of the two alkalies, potassium is by far the predominant impurity that needs close attention from the plant chemist. During the burning process, alkalies vaporize in the lower part of the burning zone, travel with the kiln gases to the rear of the kiln, and condense again at a gas temperature of around 900 C (1650 F). These alkalies react in the colder part of the kiln with sulfur dioxide, carbon dioxide, and chlorides that are contained in the kiln gases. Thus, an internal alkali cycle is created that can lead to troublesome buildup and ring formations in the kiln. In dry- and wet-process kilns, condensation of alkalies occurs in the lower end or just below the chain section whereas in preheater, semidry, and precalciner kilns this condensation takes place in the lower stages of the preheater tower or grate preheater. Alkalies are quite an intriguing problem for any kiln manager. There must not be too much of them in the clinker; they should not be recycled and allowed to accumulate in the kiln; and yet they are found in great quantities in the raw materials. To combat these problems, various means are employed to keep these alkalies under control. In wet- and dry-process plants, part or all of the kiln dust collected in the baghouse or electrostatic precipitator must be wasted. Some plants are fortunate in that they have to waste only the last section of these dust collectors, i.e., the very fine dust particles that are richest in alkalies. Some of these plants sell this potassium-rich kiln dust as fertilizer. Preheater and precalciner kilns are equipped with alkali bypass systems at the preheater tower to control this internal alkali cycle.

There are also plants that could tolerate a slightly higher alkali content in the clinker but face the trouble of large, internal alkali cycles which call for different solutions to inhibit the vaporization of alkalies.

Sulfur (SO_3) is introduced into the kiln by the raw materials and the fuel. This impurity will also vaporize to form sulfur dioxide (SO_2) at a temperature of ≈ 1000 C (1832 F) and condense in the form of sulfates within the kiln system. They readily combine with calcium to form calcium sulfates and potassium to form potassium sulfate both of which are the prime culprits for ring and buildup problems in the upper half of the kiln system. If there is a lack of alkalies present with which the sulfur dioxide would combine to form alkali-sulfates then much of the sulfur dioxide would leave the kiln system with the kiln gases.

Experience gathered by many plant operators has shown that there should be a delicate balance between the alkali and sulfur contents in the raw mix. If the molecular ratio of *alkalies-to-sulfur* is significantly below 1.0 this gives rise to calcium sulfate buildups near the kiln inlet in preheater kilns. By raising this ratio to 1.0 (in the form of adding alkali-rich raw material) some plants have been successful in reducing the frequency of calcium sulfate buildups in the kiln. Likewise, the converse has been experienced, i.e., when this ratio exceeded 1.0 by a large margin. In such cases, due to the excess of alkalies, alkali sulfate buildups occurred. In such cases, the solution would be to lower the alkali cycle within the kiln by wasting kiln dust or letting part of the kiln exit gas bypass the preheater vessels. Another solution would be to add sulfur-bearing (SO_3) raw materials to the feed to balance the excess alkalies.

Buildup problems are usually attacked by first analyzing the material of the buildup, determining its predominant compounds (in other words trying to find out what compound caused the buildup) and finally selecting a solution that would reduce formation of these deleterious compounds.

Chlorides originate primarily from the raw materials and from the coal. For proper kiln operation, plant chemists usually try to hold the total chloride content in the raw mix below 0.02%. Chlorides, too, vaporize and react with alkalies to form alkali chloride. Alkali chlorides tend to remain in the internal kiln cycle for a long time and can lead to heavy coating and ring formation in the upper part of the rotary kiln and the lower stages of the preheater. Chlorides, even in such small quantities as 0.02% in the kiln feed can become so troublesome on some preheater kilns that they are forced to operate with a bypass of up to 15% at the preheater tower.

Fluoride, although volatile like the alkalies, sulfur, and chlorides, does not participate as readily in the internal cycle as the above-mentioned compounds. Most of the fluoride leaves the kiln with the dust in the exit gases or in the clinker.

It is primarily for these impurities that the preheater kiln didn't find rapid acceptance when it was first invented 40 years ago. Buildup problems plagued these kilns to such an extent that many managers considered the almost daily kiln shutdowns for buildup removal a problem not worth dealing with. Today these problems have been predominantly overcome and the preheater kiln has rightfully taken the leadership in preferred types of kilns when new plants are constructed.

In coal-fired kilns, a plant chemist must also consider the *ash* from the combustion of coal as an ingredient in the kiln-feed mix. If a kiln fires 25 tons/h of coal having an ash content of 17%, there will be 4250 kg of coal ash added hourly to the system. Typical coal ash consists of:

SiO_2	Al_2O_3	Fe_2O_3	CaO	SO_3, Cl, K_2O, Na_2O
47%	27%	20%	3%	trace

From a plant chemist's viewpoint, this is a potent material in matters of clinker chemistry modification and must therefore be closely monitored and considered in mix calculations. Problems are magnified when the kiln switches back and forth between different types of fuels burned, e.g., when natural gas is fired part of the day and coal fired for the rest. Since natural gas does not contain the clinker modifying ash, the plant chemist must find a kiln-feed blend that is suitable for both types of fuel burning. It should be suitable from a burning as well as a quality-control viewpoint. In such instances, the resultant kiln-feed mix is a compromise at best. But, it is far better to compromise than to design a mix for coal firing and then switch to natural gas without making a mix adjustment. The repercussions of such actions can be severe because the burnability of this kiln feed is drastically changed the moment the switch to natural gas is made. In this example, the kiln feed would be much harder to burn because the fluxing alumina and iron components in the coal ash would suddenly be absent. A plant chemist of a precalciner plant must also give the same attention to the amount of ash that enters the system at the flash furnace.

Now, it will be assumed that all these factors have been given due consideration by the chemist, and that the proper mix of different raw materials to obtain a typical kiln-feed composition (Table 9.2) is found. (Note: The chemical analysis below is used as the basis for all remaining discussions in this chapter. This analysis is only an example for illustrative purposes; every plant's kiln feed will somehow deviate from the one shown below.)

TABLE 9.2

KILN FEED COMPOSITION

	Loss free (%)	Raw basis (%)
SiO_2	20.71	13.34
Al_2O_3	4.84	3.12
Fe_2O_3	2.78	1.79
CaO	67.43	43.42
MgO	2.58	1.66
Na_2O	0.22	0.14
K_2O	0.85	0.55
SO_3	0.15	0.10
Loss on ignition	0.00	35.60

The chemist now looks at the ash that enters the clinker and calculates what the theoretical clinker composition will be after the kiln feed has been burned and the ash included. Assume in this example that the clinker contains 3% coal ash:

TABLE 9.3

EFFECT OF COAL ASH ON CLINKER COMPOSITION

	Percent on loss-free basis		
	Kiln feed	ash	potential clinker
SiO_2	20.71	47.00	21.50
Al_2O_3	4.84	27.00	5.50
Fe_2O_3	2.78	20.00	3.30
CaO	67.43	3.00	65.50
MgO	2.58	0.00	2.50
SO_3	0.15	trace	0.15
Na_2O	0.22	trace	0.83
K_2O	0.85	trace	—

This clinker composition is only a preliminary potential wherein the actual composition of the clinker can vary due to the volatility of the alkalies and the sulfur. The important thing to remember is that, at this stage, there is not a material that has cementitious (hydraulic) properties. That is where the kiln comes in for it is here that this kiln feed is transformed into clinker minerals to obtain the ultimate properties of what is known as cement. During burning, this kiln feed forms the four main clinker compounds:

Tricalcium silicate	$3CaO \cdot SiO_2$	$= C_3S$ (alite)
Dicalcium silicate	$2CaO \cdot SiO_2$	$= C_2S$ (belite)
Tricalcium aluminate	$3CaO \cdot Al_2O_3$	$= C_3A$
Tetracalcium aluminoferrite	$4CaO \cdot Al_2O_3 \cdot Fe_2O_3$	$= C_4AF$

This compound composition is calculated with the help of the Bogue formulas, from the potential clinker analysis above (loss-free basis).

9.3   BOGUE FORMULAS FOR CLINKER AND CEMENT CONSTITUENTS

For a cement chemist, these formulas are the most important and frequently used indicators of the chemical properties of a cement or clinker. The constituents calculated by these formulas, however, are only the potential compositions when the clinker has been burned and cooled at given conditions. Changes in cooling rate or burning temperature can modify the true constituent composition to a considerable extent.

a)   Bogue Formulas for Cement Constituents.

If $A/F = > 0.64$

$$C_3S = 4.071CaO - (7.602SiO_2 + 6.718Al_2O_3 + 1.43\ Fe_2O_3 + 2.852SO_3)$$

$$C_2S = 2.867SiO_2 - 0.7544C_3S$$

$$C_3A = 2.65Al_2O_3 - 1.692Fe_2O_3$$

C_4AF $= 3.043Fe_2O_3$

If $A/F = < 0.64$

C_3S $= 4.071CaO - (7.602SiO_2 + 4.479Al_2O_3$
$+ 2.2859Fe_2O_3 + 2.852SO_3)$

C_2S $= 2.867SiO_2 - 0.7544C_3S$
C_3A $= 0$
$(C_4AF + C_2F)$ $= 2.1Al_2O_3 + 1.702Fe_2O_3$

b) Bogue Formulas for Clinker Constituents.

When appreciable amounts of SO_3 and Mn_2O_3 are present in the clinker, the values of the chemical analysis have to be recalculated to take into account the amount of CaO that has been combined with SO_3, the amount of free lime present and the Mn_2O_3.

The values to be used in the Bogue formulas are:

Fe_2O_3 $= Fe_2O_3 + Mn_2O_3$
CaO $= CaO - $ free $CaO - (CaO$ Combined with $SO_3)$

To find the amount of CaO that is combined with SO_3 as $CaSO_4$ proceed as follows:

Step 1. If $(K_2O/SO_3) = < 1.176$ then not all of the SO_3 is combined with K_2O as K_2SO_4

$$SO_3 \text{ in } K_2O = 0.85K_2O$$

Step 2. Calculate SO_3 residue

$$SO_3 - SO_3 \text{ (in } K_2O) = SO_3 \text{ (remaining)}$$

If $[Na_2O/SO_3 \text{ (remain.)}] = < 0.774$ then not all of the remaining SO_3 is combined with Na_2O as Na_2SO_4.

$$SO_3 \text{ in } Na_2O = 1.292Na_2O$$

Step 3. Calculate the amount of CaO that has combined with the SO_3 as $CaSO_4$

$$CaO \text{ (in } SO_3) = 0.7[SO_3 - SO_3 \text{ (in } K_2O) - SO_3 \text{ (in } Na_2O)]$$

Having determined the appropriate values for the CaO and Fe_2O_3, one can then proceed to calculating the potential clinker constituents by using the previously given Bogue formulas. When the Bogue formulas are used for kiln feed compositions, keep in mind that the coal ash addition, dust losses, and alkali cycles can alter the final composition of the clinker. Also, it is necessary to use the analysis on a "loss-free" basis in the calculations of the constituents.

Tricalcium silicate is an important constituent as it is responsible mainly for early strength development of mortar and concrete. Regular portland cement kiln feed has usually a C_3S potential of 52–62%. Kiln feed with a potential in excess of 65% is extremely difficult to burn and has a poor coating characteristic.

Dicalcium silicate accounts for approximately 22% of the clinker. Because a higher temperature is required to form C_3S than C_2S, under-burning could result in a higher content of C_2S and a lower content of C_3S.

Tricalcium aluminate is responsible for the workability of the mortar. The higher the C_3A content, the higher the plasticity (workability) of the mortar. This explains why kiln feed for the so-called plastic cements has a higher C_3A potential than that for regular cement in which the C_3A amounts to 6–8% of the clinker. Concrete containing cement high in C_3A is not as resistant to attack by sulfates in soil or water exposure as is concrete made with low C_3A cement.

Tetracalcium aluminoferrite governs the color of the cement. The higher the content of CA_4F in the clinker, the darker the cement. This is undesirable, as users almost unanimously prefer a light-colored cement. Iron has the desirable property of acting as a fluxing agent in the kiln, facilitating formation of other compounds of the cement at somewhat lower temperature than would otherwise be possible.

It is quite obvious that it is necessary to have a continuing analysis of the material going into the kiln, if there is going to be adequate control of the product coming out of the other end of the kiln. It is the responsibility of the plant chemist to determine the composition of these materials and to proportion them to produce a kiln feed that ensures a uniform, high-quality clinker, combined with good burnability. Continuously uniform composition of the kiln feed is of greatest importance for proper operation of the kiln.

Various systems are employed to introduce the feed into the kiln depending on whether the wet or dry process is to be used. These systems all serve the same purpose: to feed the kiln at a steady and uniform rate with as little fluctuation as possible, which means that each raw material must be carefully metered or measured.

Using our examples above and the appropriate Bogue formulas, the following potential clinker compound content would be obtained:

C_3S	C_2S	C_3A	C_4AF
61.8	15.0	8.73	10.04

It must be again stressed here that the clinker compounds as calculated by the Bogue formulas are only potential in nature. Formation of these depends on the temperature, time of exposure, and cooling rate of the clinker in the burning zone and in actuality are quite different than the calculated values.

Up until now, a plant chemist has laid the foundation for the possible quality of the cement that will be produced from this clinker. But, his job doesn't end here. It is his duty to reconcile this clinker with the burnability and coatability of this clinker. In other words, he not only has to concern himself with making a good quality cement but must also give due attention to the ease at which this clinker can be burned. Later in this chapter the microstructures of cement clinkers will be discussed and it will be shown that a plant chemist, in today's cement technology, should and must also concern himself with the burning-zone environments that will ultimately affect the quality of cement.

9.4   INFLUENCE OF FEED COMPOSITION ON BURNABILITY

The "burnability" of a kiln feed is the relative ease or difficulty with which the feed is changed into a clinker in the kiln; that is, it is an indication of the amount of fuel required to burn the kiln feed into a clinker of good quality. Although it is highly desirable to produce at all times the same composition of kiln feed, this cannot readily be done. One reason is that most cement plants manufacture different types of cement, such as high-early-strength, block and sulfate-resistant; therefore, composition of the kiln feed must change from time to time as different kinds of cement are being manufactured. Every time the feed composition changes, burnability in the kiln will also change.

A plant chemist, when calculating the kiln-feed composition, will employ certain formulas to ensure that the finished product meets the specifications of the type of portland cement to be made. Kiln-feed compositions are identified by a multitude of factors and indexes which are also used to express burnability. These are discussed briefly below.

Silica Ratio.

The Silica Ratio is found by dividing the silica content by the sum of the contents of alumina and iron in the kiln-feed blend. That is,

$$S/R = \frac{SiO_6}{Al_2O_3 + Fe_2O_3}$$

Increasing the silica ratio produces a clinker that is more difficult to burn; in other words, the clinker is "harder" to burn. It is mainly the content of alumina and ferric oxide that governs the combination of calcium and silica at lower sintering temperatures.

At this point, it is appropriate to define the terms "easy burning" and "hard burning." In this text, an easy-burning kiln feed is one that requires less fuel to burn to a clinker than a hard-burning feed.

Alumina-Iron Ratio.

The Alumina-Iron Ratio is found by dividing the alumina content of the kiln feed by the iron content. That is,

$$A/F = \frac{Al_2O_3}{Fe_2O_3}$$

The higher the ratio, the harder the burning. Iron has a favorable influence on the speed of reaction between lime and silica; therefore, one can also say: Other values remaining constant, a higher iron content leads to easier burning. Because both the numerator and denominator in the equation are expressions of fluxing components, however, the alumina alone is not used to express burnability.

This ratio indicates the quantity of initial liquid phase present during burning. It is generally accepted that an A/F ratio between 1.4–1.6 is a desirable optimum level and most beneficial to the burning of the clinker. The higher this ratio, the harder the clinker will be to burn.

The Lime-Saturation Factor.

This factor has been used for kiln-feed control for many years in Europe and only recently has also found acceptance by American cement manufacturers. When the lime-saturation factor approaches unity, the clinker is difficult to burn and often shows excessively high free-lime contents. A clinker, showing a lime-saturation factor of 0.97 or higher approaches the threshold of being "overlimed" wherein the free-lime content could remain at high levels regardless of how much more fuel the kiln operator is feeding to the kiln.

If $A/F > 0.64$

$$LSF = \frac{CaO}{2.8SiO_2 + 1.65Al_2O_3 + 0.35Fe_2O_3}$$

If $A/F < 0.64$

$$LSF = \frac{CaO}{2.8SiO_2 + 1.1Al_2O_3 + 0.7Fe_2O_3}$$

This lime-saturation factor when viewed in context with other indicators is an excellent indicator of what the free-lime content will be in the clinker when it has been burned at normal temperatures.

There have been instances in the past where a foreman, upon learning from the lab results that the free-lime content was too high, approached the kiln burner and asked him to burn the kiln hotter to lower the free lime. It is true that lower burning zone temperatures deliver higher and, conversely, "hotter" kilns resulting in lower free lime in the clinker. But, this is not the only factor. High free lime can be associatd with too low a burning-zone temperature *only* when the feed mix and the fuel burned remain unchanged, and when the lime-saturation factor of the feed is below the so-called saturation point. If for whatever reason the mix should suddenly show a lime-saturation factor of, e.g., 0.97 or higher, it would be very difficult for an operator to lower the resultant free lime by raising the burning-zone temperature. Such action most likely would do more harm to the coating and refractory than it would do any good to the clinker quality.

The opposite has also been observed where complaints were voiced to the kiln operator about burning the kiln too hot. The reason given was that the free lime was consistently too low in the clinker. Again, a badly underlimed mix, having a lime saturation of less than 0.88, tends to deliver clinker that is low on tree lime.

The point to remember is that when the free lime in the clinker is not up to standards, a check with the laboratory should be made first to see if the mix (lime-saturation factor) has changed. If this factor is still within normal ranges, then and only then, is there an indication that the kiln operator might not have burned the clinker at the proper temperature.

The permissible range of variation for free-lime contents varies among different plants but the majority of the plants attempts to obtain values that are between 0.4 and 1.2%. Experience has shown that when the clinker is burned as close as possible to the 0.8% level of free lime, the mix is then within acceptable levels.

The Hydraulic Ratio.

The *hydraulic ratio*, developed by W. Michaelis over 100 years ago, is very seldom used any more in modern cement technology for kiln-feed control but is here included for plants that still regard this ratio as significant:

$$HR = \frac{CaO}{SiO_2 + Al_2O_3 + Fe_2O_3}$$

Percent Liquid.

Clinker, when burned at 1450 C (2642 F) will be in a so-called semiliquid state. This viscous appearance of the clinker bed is a very important control factor for a kiln operator when viewing the burning zone. This will be discussed in greater detail later on. The *percent liquid* is calculated by the Lea and Parker formula as follows:

percent liquid @ 1450 C = $3.0Al_2O_3 + 2.25Fe_2O_3 + MgO$ + alkalies

or by the formula which gives the same results:

percent liquid @ 1450 C = $1.13C_3A + 1.35C_4AF + MgO$ + alkalies

In both these formulas, the restriction applies that the MgO content is limited to a maximum of 2%. In other words, a value of not more than 2% MgO can be used in these formulas.

Most portland cement clinkers show a liquid content of 25–27.5%. Higher liquids produce stickier burning-zone clinker-bed appearances. Since the percent liquid as calculated by the above formulas applies to a temperature of 1450 C, higher temperatures give higher liquid and, conversely, lower temperatures result in lower contents of liquid in the clinker. Also, since alumina, iron, magnesia, and alkalies are fluxes, higher liquid contents make a clinker easier to burn.

Burnability Index.

Kuehl's burnability index is based on the potential clinker compounds C_3S, C_4AF, and C_3A. The higher the content of C_3S with corresponding lower contents in C_4AF or C_3A, the harder the clinker is to burn.

$$Burnability\ Index\ = \frac{C_3S}{C_4AF + C_3A}$$

The Burnability Factor.

This factor was first introduced by this author in the original *Rotary Cement Kiln* book. It was brought to the attention of the author that several chemists and engineers have made further research on the application potential of combining the lime-saturation factor with the silica ratio to express burnability. As mentioned in the original writeup, this formula was developed based on pure empirical notions and observations and, hence, was suspect in its fundamental reasoning.

Burnability Factor = $100(LSF) + 10(S/R) — 3(MgO$ + alkalies)

The laboratories of F.L. Smidth, Copenhagen, have recently presented their results of an investigation on this subject. Their findings are significant since they used scientific methods to arrive at a direct indicator of percent free lime in the clinker when the clinker is burned to 1500 C. The effect of alkalies and magnesia on burnability have been assumed constant in their formula, but due consideration was given to the effect of kiln-feed fineness on burnability. The FLS formula is:

$$\% \text{ free lime @ } 1500 \text{ C} = [(0.33LSF) + (1.8S/R)] - 34.98 + 0.5a = 0.13b$$

where:
   a = % retained on 45μ sieve (after acidification)
   b = % retained on 125μ sieve (after acidification)

Other chemists have used the *Peray* burnability factor for mix control but have determined that the coefficients in the formula had to be changed. In other words:

$$Burnability\ Factor = x(LSF) + y(S/R) - z(MgO + \text{alkalies})$$

where x, y, and z were determined by multiple regression analysis.

There have been others that have pursued this question of expressing burnability by using a meaningful index or factor.

Analysis of Burnability.

In addition to the kiln-feed composition discussed above, the operator of a wet-process kiln has to consider also the moisture content of the kiln feed, as this indirectly affects burnability of the clinker. With unchanged composition of the kiln feed, a higher moisture content results in easier burning. The reason for the change in burnability is the simple fact that less feed is available to be burned when the moisture content is higher.

Because changes in kiln-feed composition have a large influence on kiln operation, it is important that the kiln operator be advised by the laboratory well in advance of any upcoming change in composition. Another good procedure is to note the chemical characteristics of the kiln feed every day in the kiln log.

So far only the influence of chemical properties on burnability in the kiln feed has been discussed. Plant operators must also pay attention to the kiln-feed fineness as these physical properties of the feed can influence burnability and the stability of the kiln operation. Since each plant produces clinker by using different raw materials and various types of kilns are in use, there are no clear cut standards in matters of kiln-feed fineness that would apply to all kilns. There is, however, a consensus among operators that:

- The coarse fractions in the kiln feed tend to be more significant than the finer fractions in their relationship to burnability. Hence, each plant needs to test for and specify the maximum limits of allowable fractions retained on the 30- or 50-mesh sieve (300, 500µ sizes respectively).
- The kiln feed has to be ground consistently uniform and with as little variations in particle-size distribution as possible on a daily basis.

It is essential to make a clear distinction between a feed blend that tends to give a better visibility in the burning zone and a blend that requires less fuel to burn. Certain feed compositions create "dirty" conditions in the burning zone. Here a kiln operator could come to the wrong conclusion that because of the poor visibility the burning zone has cooled down. The action of raising the fuel rate in order to clear the burning zone could result in an overburned clinker.

On the other hand, another kiln-feed blend could possibly improve visibility in the burning zone. In such a case, a kiln operator could wrongly reduce the fuel rate because he concluded that the burning zone was warming up. The final result of his action could then be an underburned clinker.

In Tables 9.4–9.7 the original potential clinker composition and the various ratios and factors are computed. The percent free lime at 1500 C is calculated using the F.L. Smidth formula but assuming that the kiln-feed fineness (45 and 125µ respectively) remains constant at 0.04 and 0.11 percent respectively.

For the purpose of illustration, changes of identical magnitude have been made in all four main oxides to show the reader how changes affect these various factors.

Several of the computed compositions are completely unacceptable in terms of burnability and clinker quality but they serve the purpose of familiarizing the reader with some of the intricate aspects of quality control.

TABLE 9.4

CLINKER PROPERTY CHANGES (when basic oxides are varied by 0.25%)

SiO$_2$	21.50	21.25	21.25	21.25	21.75	21.75	21.75	21.50	21.50	21.50	21.50	21.50	21.50
Al$_2$O$_3$	5.40	5.65	5.40	5.40	5.15	5.40	5.40	5.15	5.15	5.65	5.65	5.40	5.40
Fe$_2$O$_3$	3.30	3.30	3.55	3.30	3.30	3.05	3.30	3.55	3.30	3.05	3.30	3.55	3.05
CaO	65.50	65.50	65.50	65.75	65.50	65.50	65.25	65.50	65.75	65.50	65.25	65.25	65.75
MgO	2.50	2.50	2.50	2.50	2.50	2.50	2.50	2.50	2.50	2.50	2.50	2.50	2.50
Alkali as Na$_2$O	0.83	0.83	0.83	0.83	0.83	0.83	0.83	0.83	0.83	0.83	0.83	0.83	0.83
SO$_3$	0.15	0.15	0.15	0.15	0.15	0.15	0.15	0.15	0.15	0.15	0.15	0.15	0.15
Ignition loss	0.40	0.40	0.40	0.40	0.40	0.40	0.40	0.40	0.40	0.40	0.40	0.40	0.40
TOTAL	99.58	99.58	99.58	99.58	99.58	99.58	99.58	99.58	99.58	99.58	99.58	99.58	99.58
C$_3$S	61.78	62.00	63.33	64.70	61.56	60.24	58.87	63.11	64.48	60.46	59.09	60.41	63.16
C$_2$S	15.03	14.15	13.15	12.11	15.91	16.91	17.95	14.03	13.00	16.03	17.07	16.07	13.99
C$_3$A	8.73	9.39	8.30	8.73	8.06	9.15	8.73	7.64	8.06	9.81	9.39	8.30	9.15
C$_4$AF	10.04	10.04	10.80	10.04	10.04	9.28	10.04	10.80	10.04	9.28	10.04	10.80	9.28
Silica ratio	2.47	2.37	2.37	2.44	2.57	2.57	2.50	2.47	2.54	2.47	2.40	2.40	2.54
Lime saturation	93.22	93.60	94.04	94.52	92.84	92.41	91.95	93.65	94.13	92.79	92.32	92.75	93.69
A/F ratio	1.64	1.71	1.52	1.64	1.56	1.77	1.64	1.45	1.56	1.85	1.71	1.52	1.77
Liquid (%)	26.25	27.00	26.80	26.25	25.50	25.70	26.25	26.05	25.50	26.45	27.00	26.80	25.70
Hydraulic ratio	2.17	2.17	2.17	2.20	2.17	2.17	2.14	2.17	2.20	2.17	2.14	2.14	2.20
Lea burnability index	3.29	3.19	3.31	3.45	3.40	3.27	3.14	3.42	3.56	3.17	3.04	3.16	3.43
Peray burnability factor	107.90	107.40	107.80	109.00	108.60	108.20	107.00	108.40	109.60	107.50	106.40	106.80	109.10
FLS % free lime	0.26	0.22	0.36	0.64	0.32	0.18	0.00	0.41	0.70	0.12	0.00	0.00	0.55

NOTE: FLS % free lime calculated on assumption that kiln-feed fineness factor remains constant at 0.0343.

TABLE 9.5

CLINKER PROPERTY CHANGES (when basic oxides are varied by 0.50%)

SiO_2	21.50	21.00	21.00	21.00	22.00	22.00	22.00	21.50	21.50	21.50	21.50	21.50	21.50
Al_2O_3	5.40	5.90	5.40	5.40	4.90	5.40	5.40	4.90	4.90	5.90	5.90	5.40	5.40
Fe_2O_3	3.30	3.30	3.80	3.30	3.30	2.80	3.30	3.80	3.30	2.80	3.30	3.80	2.80
CaO	65.50	65.50	65.50	66.00	65.50	65.50	65.00	65.50	66.00	65.50	65.00	65.00	66.00
MgO	2.50	2.50	2.50	2.50	2.50	2.50	2.50	2.50	2.50	2.50	2.50	2.50	2.50
Alkali as Na_2O	0.83	0.83	0.83	0.83	0.83	0.83	0.83	0.83	0.83	0.83	0.83	0.83	0.83
SO_3	0.15	0.15	0.15	0.15	0.15	0.15	0.15	0.15	0.15	0.15	0.15	0.15	0.15
Ignition loss	0.40	0.40	0.40	0.40	0.40	0.40	0.40	0.40	0.40	0.40	0.40	0.40	0.40
TOTAL	99.58	99.58	99.58	99.58	99.58	99.58	99.58	99.58	99.58	99.58	99.58	99.58	99.58
C_3S	61.78	62.23	64.87	67.62	61.34	58.70	55.95	64.43	67.18	59.14	56.39	59.03	64.53
C_2S	15.03	13.26	11.27	9.19	16.80	18.79	20.87	13.04	10.96	17.03	19.10	17.11	12.96
C_3A	8.73	10.05	7.88	8.73	7.40	9.57	8.73	6.56	7.40	10.90	10.05	7.88	9.57
C_4AF	10.04	10.04	11.56	10.04	10.04	8.52	10.04	11.56	10.04	8.52	10.04	11.56	8.52
Silica ratio	2.47	2.28	2.28	2.41	2.68	2.68	2.53	2.47	2.62	2.47	2.34	2.34	2.62
Lime saturation	93.22	93.99	94.87	95.84	92.46	91.62	90.70	94.09	95.05	92.36	91.43	92.28	94.16
A/F ratio	1.64	1.79	1.42	1.64	1.48	1.93	1.64	1.29	1.48	2.11	1.79	1.42	1.93
Liquid (%)	26.25	27.74	27.35	26.25	24.75	25.15	26.25	25.85	24.75	26.65	27.74	27.35	25.15
Hydraulic ratio	2.17	2.17	2.17	2.22	2.17	2.17	2.12	2.17	2.22	2.17	2.12	2.12	2.22
Lea burnability index	3.29	3.10	3.34	3.60	3.52	3.24	2.98	3.56	3.85	3.05	2.81	3.04	3.57
Peray burnability factor	107.90	106.80	107.70	110.00	109.30	108.50	106.00	108.80	111.30	107.10	104.80	105.70	110.40
FLS % free lime	0.26	0.18	0.47	1.03	0.40	0.12	0.00	0.55	1.14	0.00	0.00	0.00	0.85

NOTE: FLS % free lime calculated on assumption that kiln-feed fineness factor remains constant at 0.0343.

TABLE 9.6

CLINKER PROPERTY CHANGES (when basic oxides are varied by 0.75%)

SiO$_2$	21.50	20.75	20.75	20.75	22.25	22.25	22.25	21.50	21.50	21.50	21.50	21.50	21.50
Al$_2$O$_3$	5.40	6.15	5.40	5.40	4.65	5.40	5.40	4.65	4.65	6.15	6.15	5.40	5.40
Fe$_2$O$_3$	3.30	3.30	4.05	3.30	3.30	2.55	3.30	4.05	3.30	2.55	3.30	4.05	2.55
CaO	65.50	65.50	65.50	66.25	65.50	65.50	64.75	65.50	66.25	65.50	64.75	64.75	66.25
MgO	2.50	2.50	2.50	2.50	2.50	2.50	2.50	2.50	2.50	2.50	2.50	2.50	2.50
Alkali as Na$_2$O	0.83	0.83	0.83	0.83	0.83	0.83	0.83	0.83	0.83	0.83	0.83	0.83	0.83
SO$_3$	0.15	0.15	0.15	0.15	0.15	0.15	0.15	0.15	0.15	0.15	0.15	0.15	0.15
Ignition loss	0.40	0.40	0.40	0.40	0.40	0.40	0.40	0.40	0.40	0.40	0.40	0.40	0.40
TOTAL	99.58	99.58	99.58	99.58	99.58	99.58	99.58	99.58	99.58	99.58	99.58	99.58	99.58
C$_3$S	61.78	62.45	66.41	70.54	61.12	57.15	53.03	65.75	69.88	57.82	53.69	57.66	65.91
C$_2$S	15.03	12.38	9.39	6.28	17.68	20.67	23.79	12.04	8.93	18.02	21.14	18.14	11.92
C$_3$A	8.73	10.71	7.46	8.73	6.74	10.00	8.73	5.47	6.74	11.98	10.71	7.46	10.00
C$_4$AF	10.04	10.04	12.32	10.04	10.04	7.76	10.04	12.32	10.04	7.76	10.04	12.32	7.76
Silica ratio	2.47	2.20	2.20	2.39	2.80	2.80	2.56	2.47	2.70	2.47	2.28	2.28	2.70
Lime saturation	93.22	94.38	95.72	97.19	92.09	90.84	89.48	94.53	95.98	91.94	90.56	91.81	94.64
A/F ratio	1.64	1.86	1.33	1.64	1.41	2.12	1.64	1.15	1.41	2.41	1.86	1.33	2.12
Liquid (%)	26.25	28.49	27.89	26.25	24.00	24.60	26.25	25.65	24.00	26.85	28.49	27.89	24.60
Hydraulic ratio	2.17	2.17	2.17	2.25	2.17	2.17	2.09	2.17	2.25	2.17	2.09	2.09	2.25
Lea burnability index	3.29	3.01	3.36	3.76	3.64	3.22	2.83	3.70	4.16	2.93	2.59	2.91	3.71
Peray burnability factor	107.90	106.30	107.70	111.10	110.10	108.80	105.10	109.30	113.00	106.70	103.30	104.60	111.70
FLS % free lime	0.26	0.15	0.59	1.42	0.48	0.07	0.00	0.70	1.59	0.00	0.00	0.00	1.15

NOTE: FLS % free lime calculated on assumption that kiln-feed fineness factor remains constant at 0.0343.

TABLE 9.7

CLINKER PROPERTY CHANGES (when basic oxides are varied by 1.00%)

SiO$_2$	21.50	20.50	20.50	20.50	22.50	22.50	22.50	21.50	21.50	21.50	21.50	21.50	21.50
Al$_2$O$_3$	5.40	6.40	5.40	5.40	4.40	5.40	5.40	4.40	4.40	6.40	6.40	5.40	5.40
Fe$_2$O$_3$	3.30	3.30	4.30	3.30	3.30	2.30	3.30	3.30	4.30	3.30	2.30	4.30	2.30
CaO	65.50	65.50	65.50	66.50	65.50	65.50	64.50	66.50	65.50	64.50	65.50	64.50	66.50
MgO	2.50	2.50	2.50	2.50	2.50	2.50	2.50	2.50	2.50	2.50	2.50	2.50	2.50
Alkali as Na$_2$O	0.83	0.83	0.83	0.83	0.83	0.83	0.83	0.83	0.83	0.83	0.83	0.83	0.83
SO$_3$	0.15	0.15	0.15	0.15	0.15	0.15	0.15	0.15	0.15	0.15	0.15	0.15	0.15
Ignition loss	0.40	0.40	0.40	0.40	0.40	0.40	0.40	0.40	0.40	0.40	0.40	0.40	0.40
TOTAL	99.58	99.58	99.58	99.58	99.58	99.58	99.58	99.58	99.58	99.58	99.58	99.58	99.58
C$_3$S	61.78	62.67	67.96	73.46	60.90	55.61	50.11	67.07	72.57	56.50	50.99	56.28	67.28
C$_2$S	15.03	11.50	7.51	3.36	18.56	22.55	26.70	11.04	6.89	19.02	23.17	19.18	10.88
C$_3$A	8.73	11.38	7.03	8.73	6.08	10.42	8.73	4.38	6.08	13.07	11.38	7.03	10.42
C$_4$AF	10.04	10.04	13.08	10.04	10.04	7.00	10.04	13.08	10.04	7.00	10.04	13.08	7.00
Silica ratio	2.47	2.11	2.11	2.36	2.92	2.92	2.59	2.47	2.79	2.47	2.22	2.22	2.79
Lime saturation	93.22	94.77	96.59	98.57	91.72	90.08	88.28	94.98	96.92	91.53	89.69	91.34	95.12
A/F ratio	1.64	1.94	1.26	1.64	1.33	2.35	1.64	1.02	1.33	2.78	1.94	1.26	2.35
Liquid (%)	26.25	29.24	28.44	26.25	23.25	24.05	26.25	25.45	23.25	27.05	29.24	28.44	24.05
Hydraulic ratio	2.17	2.17	2.17	2.28	2.17	2.17	2.07	2.17	2.28	2.17	2.07	2.07	2.28
Lea burnability index	3.29	2.93	3.38	3.91	3.78	3.19	2.67	3.84	4.50	2.82	2.38	2.80	3.86
Peray burnability factor	107.90	105.90	107.70	112.10	110.90	109.30	104.10	109.70	114.80	106.20	101.90	103.50	113.00
FLS % free lime	0.26	0.13	0.73	1.82	0.58	0.04	0.00	0.84	2.06	0.00	0.00	0.00	1.47

NOTE: FLS % free lime calculated on assumption that kiln-feed fineness factor remains constant at 0.0343.

Kiln-feed blends for production of special cement-clinker types such as high-early, sulfate-resistant, low-alkali, and oilwell cement will all have somehow different properties and chemical compositions. Changing from one type of clinker to another burn always requires special attention from the kiln operator and advice of such a change should be given well before this new feed is being used in the kiln.

The Liter-Weight Test.

One of the easiest tests a kiln operator can perform to learn if he has burned the clinker at the proper temperature is the liter-weight test. Free-lime content also gives essentially the same information but analysis for free-lime content takes up to an hour until the results are reported. Since the sample is usually taken from the outlet of the cooler, the results tell what was done 1.5–2 h before which, from an operator's viewpoint, is not much help. In the liter-weight test, the sample is first passed through a 10-mm screen then through a 5-mm screen. The fraction retained on the 5-mm screen is then allowed to fall through a prescribed distance into a 1000-ml container which is in the shape of a frustrum of a cone with the small end up. The weight of the clinker in this container, called the liter weight, indicates how well the clinker has been burned, as a hard-burned clinker has a higher liter weight than a soft-burned clinker, provided there is no change in raw-mix composition. A well-burned clinker has a liter weight between 1250–1350 g. The liter weight can vary considerably, even though the clinker is well burned, between one feed composition and another. If the raw-feed composition remains constant, the best clinker will have the appropriate liter weight and lowest free-lime content. The time required to run a liter-weight test is approximately 5 min.

Care must be taken to assure that the entire test procedure is carried out in the same consistent manner. From personal experience it better serves the operator to determine at what time the liter weight should be performed. In a well-running kiln, there is not much use in wasting time running a liter-weight test every hour. Liter-weight tests should be done when the operator or foreman is not quite sure if the clinker is being burned at the proper temperature.

Here, too, as with the free-lime test, there is the disadvantage of the time lag to consider since most samples are extracted from the cooler discharge. Some plants have overcome this by extracting the sample directly from the kiln hood before it falls into the cooler using careful cooling procedures to guard against burns.

How many and what types of clinker indexes, factors, etc., are to be used for effective quality control is a matter that has to be decided by the operator and plant chemist. This author has used a backward approach to arrive at an optimum mix that theoretically would successfully produce a desired clinker. First, certain fixed desired properties like the lime-saturation factor, silica ratio, and A/F ratio are set and using these as constants the needed oxide composition is arrived at. This method of optimum-mix design is very tedious when done by hand since it involves repeated trial-and-error calculations until the right suitable mix is reached. But, with the help of a computer program, this work is greatly simplified and quickly accomplished. An example using this procedure is shown in Table 9.8.

In the preceding pages the chemistry of the kiln feed and clinker have been extensively discussed. The novice reader should now have a fairly broad knowledge of the many problems and factors that are associated with making cement clinker in a kiln. For new kiln operators there will come a time when the realization that something is not quite right with the mix occurs. It might suddenly happen while burning a problematic kiln feed. Sometimes these problems can persist for several days and a kiln operator can become frustrated. It is common for kiln operators, whenever the kiln doesn't operate and handle properly, to blame these problems mistakenly on the laboratory staff. In most instances this is not justified. It must be realized that the plant chemists face just as many obstacles as the kiln operator. Most of the time a chemist is aware that the mix is not up to his liking but he can do nothing about it because of many factors not directly under his control. Raw materials might not be available to make appropriate corrections, kiln-feed reserves could be at low levels, or the raw-grinding department might face it's own disturbances.

In other instances the laboratory might get right back at the kiln operator and tell him that there is nothing wrong with the mix. The best way to guard against this type of stalemate is to establish a good dialogue between the laboratory and the kiln control room. A kiln operator can accept problems when they are explained. Being made aware of potential upcoming disturbances in the kiln is a more tolerable situation.

Assuming that the point has been reached where a shining clinker of superior quality has discharged from the cooler, the stage would be set for a good finished product. However, the cement as yet has not been produced. If just clinker alone is ground, the final product would be an inferior, unworkable cement. To control setting time, gypsum has to be added to the clinker during the finish-grinding process. Typical portland cement contains approximately 5% of gypsum.

TABLE 9.8

OPTIMUM CLINKER COMPOSITION

ENTER LOSS-FREE VALUES (typical in clinker)

Alkalies (K_2O & Na_2O)		0.60
Sulfur (SO_3)		0.40
Magnesia (MgO)		2.30
Iron (Fe_2O_3) most desirable percent	$a =$	3.15

Step 1. Set desired A/F ratio $b =$ 1.80 →

Step 2. Calculate Al_2O_3 content $= a*b$ $c =$ 5.67   OK? Y

Step 3. Set desired silica ratio $d =$ 2.45 →

Step 4. Calculate SiO_2 content $= d*(a+c)$ $e =$ 21.61   OK? Y

Step 5. Set the desired Lime-Saturation factor (Kuehl) $f =$ 0.93 →

Step 6. Calculate CaO content
$= f*[2.8e) + (1.65c) + (0.35a)]$ $g =$ 66.00 →

Step 7. Check sum total all oxides (must be > 99.5 < 100.0)   99.72   OK? Y

Step 8. Calculate potential clinker compounds by Bogue

(when $A/F > 0.64$)		
C_3S	60.66	OK?
C_2S	16.19	OK?
C_3A	9.70	OK?
C_4AF	9.59	

Step 9. Calculate burnability factor   108.80   OK?

(Proceed with calculating coal-ash addition effect before determining the actual kiln-feed mix composition.)

At this stage, the production of cement, the binder that holds the sand and the aggregates in the concrete and mortar together, is completed. This is the most durable construction material that is known. No wood, glass, or steel construction will ever have the life span of concrete. The many things that were made of clinker serve as tributes to the kiln operator. An old, now retired kiln operator, told this author a few years ago that one of the most memorable things he did in life was to be involved in making the cement used for the construction of the Hoover Dam in Nevada. Considered at that time to be one of the eight technical wonders of the world, it is still around and doing well. What a monument to a kiln operator that had no automatic controller, no computers, no air-conditioned control room, and operated a kiln that was driven by a leather belt around the shell.

9.5   CLINKER MICROSCOPY

It was mentioned earlier that the clinker very rarely contains the amounts of clinker constituents as are calculated by the Bogue formulas. Sooner or later, every plant chemist also faces the situation where a drop in cement quality cannot be explained by the earlier mentioned clinker indexes, the chemical composition of the clinker, or by factors related to the finish grinding of the cement. In many respects the plant chemist can have the same problems as the kiln operator in that once in a while things go completely contrary to what one has learned or experienced over a long period of time. In short, poor cement quality would seem to contradict apparently good clinker quality and good finish-grinding parameters. Such reasoning has led to the method of investigating the microstructures of clinker in an attempt to identify changes in clinker quality as a result of changes in the burning and cooling-zone conditions in the kiln. Examination by microscope can reveal things that chemical analysis of a clinker sample will not.

Dr. Yoshio Ono of the Onada Cement Company, Japan introduced a new technique in the early 1970's that allowed for microscopic observations of clinkers and related these to the burning conditions in a kiln and the prediction of the compressive strength of the finished cement. These observations are, to this date, qualitative and not quantitative in nature. As with any new method, the cement industry took a long time to accept it as a valuable tool in quality control. Today there still exists some limited disagreements as to the accuracy of the values found by this method in relation to observation of the burning conditions in the kiln. What most chemists and kiln operators, however, seem to agree upon is that microscopic observations of the clinker microstructure is indeed a valuable tool for kiln control and that only the interpretation of the results is in need of further study and refinement. With the passage of time, when more clinker investigations have been performed by this method, there is a good chance that the microscope might become part of standard testing equipment for clinker quality control. Several cement plants are using these microscopes on a regular basis and some have even gone so far as to locate them, for the operator's use, in the kiln control room.

As previously discussed, C_3S (alite) is the most important clinker constituent that governs the strength of the cement. Alite forms in the burning zone by combining in the liquid phase the C_2S (belite) with available free lime. Thus, it is important to burn the kiln in such a manner as to obtain the maximum amount of alite and the minimum amount of free lime in the clinker.

As this reaction proceeds, the alite crystals grow in size as the belite crystals progressively shrink and the free lime starts to disappear. Under certain burning zone and cooling conditions this newly formed alite can be unstable and revert back to C_2S (belite) and free lime in accordance with the following:

$$C_2S + \text{free } CaO + \text{(liquid phase)} \rightarrow C_3S + \text{(slow cooling)} \rightarrow C_2S + CaO$$

On slow cooling, lime is extracted from the surface of the alite crystal and leaves a belite layer (in petrographic terms this is referred to as *birefringence*) around the alite.

Hard burning of the clinker as well as too long an exposure to very high temperatures (long burning zones) leads to an increased liquid phase, causing formation of excessively large alite crystals which in turn is detrimental to the quality (strength) of the cement. Conversely, insufficient temperature leads to formation of insufficient and smaller C_3S (alite) crystals, excessive belite and high free lime which, too, leads to inferior cement quality.

Crystal sizes are therefore important to cement quality and only microscopic (petrographic) investigations of clinkers shed light on this aspect of clinker properties.

Ono's method calls for investigation of clinkers for the following variables:

a) Alite size
b) Belite size
c) Birefringence
d) Belite color

The methods of sample preparation, evaluation, and interpretation of the observations are beyond the scope of this book. Special training to this end is required. Ono has developed graphs where the observations of the above variables can be related to the burning and cooling conditions in the kiln. Users of this technique report good correlations when the kiln feed is fairly uniform but there seem to be some problems in obtaining good and meaningful results when kiln-feed chemistry and fineness fluctuate frequently.

By means of the microscope some plants have been able to identify the true causes for the formation of so-called "chestnut" clinker. For years the consensus has been that these clinker balls, having a brown center core, were caused by reducing conditions (lack of oxygen) during the sintering process wherein the trivalent iron (Fe^{III}) was transformed into ferrous oxide (Fe^{II}). Work done by G.R. Long, Blue Circle Research Division (Proceedings of the 4th International Conference on Cement Microscopy) has shown that this light brownish core in a clinker can also be caused by sintering in a normal oxidizing atmosphere with rapid cooling thereafter. The underlying causes can be differentiated from each other by visual observation. If reducing conditions cause the brownish core, there is a clearly noticeable yellow band which separates the brown inner core from the outer grey-black coat. In "chestnut" clinkers that are burned nevertheless in oxidizing environments but quick-quenched, no such yellow band has been observed.

There is no doubt that microscopic investigations of clinker have just started to come to the forefront. Application possibilities for quality control by this method are a natural step in the continuing striving by chemists and operators to improve and refine their procedures and to learn more about the complexities of cement manufacturing.

10.

Reaction Zones in the Rotary Kiln

Controlling the kiln operation means keeping a watch on all the different zones and reactions that take place within the kiln system. The chemical reactions and phase formations taking place within the kiln are discussed in Chapter 7. None of these zones can be neglected or ignored if the kiln is to be operated properly. There are five distinct zones within the kiln, their location and length being different for each type of kiln system used. These zones are shown in Table 10.1.

TABLE 10.1

ZONES IN ROTARY CEMENT KILNS

	Temperature range of material	
	°C	°F
Drying and preheating zone	15–805	60–1480
Calcining zone	805–1200	1480–2192
Upper-transition zone	1200–1400	2192–2552
Sintering zone	1400–1510	2552–2750
Cooling (lower-transition) zone	1510–1290	2750–2350

Kiln operators define the upper-transition, the sintering, and the lower-transition zones as one complete unit, referred to as the *burning zone*. The burning zone is that area in the kiln where clinker coating exists on the refractory surface.

141

In the *drying and preheating zone* all the free water in the feed is removed and the temperature of the feed is raised to approximately 805 C (1481 F) at which temperature calcination begins. During drying, the feed temperature remains at 100 C (212 F) until all the free moisture is removed. Starting at a temperature of 550 C (1022 F) the chemically combined water in the clay minerals is also driven off. Thus, in this particular zone, the reactions taking place in succession are: raising the feed temperature to the evaporation temperature, evaporating the free water, liberating the chemically combined water, and raising the feed to the calcining temperature.

From an operator's viewpoint, the *calcining zone* is the most important zone that governs subsequent events in the burning zone. Calcination is the action through which carbon dioxide is driven off from limestone and magnesium carbonate, leaving free lime and magnesia:

$$CaCO_3 \rightarrow CaO + CO_2$$
(limestone)   (free lime)   (carbon dioxide)

$$MgCO_3 \rightarrow MgO + CO_2$$
(magnesium carbonate)   (magnesia)   (carbon dioxide)

Carbon dioxide dissociates from the feed and is carried away by the kiln gases. Kiln feed that is not completely calcined before it enters the burning zone is difficult to burn and is one of the main reasons for upset kiln conditions. Thus, complete calcination of the kiln feed before it enters the burning zone is essential to proper burning of the clinker.

From the above reaction of limestone it follows that the formation of free lime in the calcining zone is accompanied by a corresponding reduction in the $CaCO_3$ (calcium carbonate) content in the feed. If samples were extracted in incremental distances along the length of the calcining zone, the rate of reaction taking place could be determined with reasonable accuracy as the feed travels down the kiln. Stopping a kiln and extracting these samples from the feed bed, however, would not serve this purpose because the feed bed would continue to be calcined after the shutdown until the temperature of the feed fell below 805 C (1481 F). Research work has been done along this line by installing sample ports on the kiln shell and sampling the feed bed while the kiln remains in operation. The results obtained are revealing and of interest to the cement industry but such procedures are prohibitive, because of the high costs for the day-to-day control of the kiln operation.

On preheater and precalciner kilns this method of checking the calcining rate is not a problem because it is easy to sample the material as it leaves individual cyclone stages. Most dry-process kilns also have a sample port at the discharge end of the chain section that allows for analysis of the degree of calcination. In this case, however, the results are not as revealing because the sample point is at a location where calcination is usually just starting. Nevertheless, checking the calcination rate on a weekly or monthly basis for dry kilns and on a daily basis for precalciner and preheater kilns is a recommended procedure.

For a continuous indication of the calcining conditions while the kiln is in operation, the operator monitors the gas analyzer that records, among other gas components, the CO_2 content of the kiln exit gases. Because it is an important control function, CO_2 control is discussed in more detail in the chapter on kiln controls.

The term *upper-transition zone* is relatively new to kiln nomenclature. In the chapter on heat transfer, mention is made of the interim-phase formations once the feed has been calcined. There is an area in the kiln where calcination and interim-phase formations overlap each other. Some of the finer particles, or the ones that were most favorably exposed to the hot kiln gases on the feed-bed surface, can reach complete calcination and start to react with the alumina, silica, or iron to form these interim compounds. With other particles, calcination would not be finished and thus no liquid formation would begin. Engineers generally refer to this area as the upper-transition zone. This is identified as the upper third of the area where clinker coating is formed in the kiln. This area is located inside the kiln, directly behind the flame where the feed bed has a "dark" color. In the temperature diagrams shown in the chapter on heat transfer, this area is identified by the sudden rapid rise in material temperature just at the start of the burning zone.

The *sintering zone* is where the final stages of clinker compound formation takes place. Unstable interim phases such as C_5A_3 and C_2F transform into the stable compounds C_3A and C_4AF. Also, C_2S combines with more free lime to form the all-important compound C_3S. sintering in the heart of the burning zone is an exothermic process, i.e., heat is evolved instead of being absorbed during this process. Sintering takes place in the middle part of the burning zone directly under the illuminated part of the flame. It is here where the most viscous behavior of the clinker bed can be observed and where the material temperatures are the highest, i.e., > 1400 C (2552 F).

The *cooling* (or *lower-transition*) *zone* entails, usually with few exceptions, the last 3–6 m (10–20 ft) of the discharge end of the kiln. Contrary to popular misconceptions, the process of clinker formation is not complete when the material has passed the hottest area. The manner in which the clinker is cooled greatly affects its quality. Depending on the location of the flame, cooling, i.e., solidification of the partly liquid compounds, can be either slow or rapid. It is generally true that rapid cooling is beneficial to both the quality of the clinker and its grindability. For rapid cooling, the sintering zone is very close to the discharge end; thus the clinker discharges from the kiln at a temperature of about 1370 C (2498 F). As soon as it enters the first quench compartment in the cooler it is rapidly cooled. Conversely, when the burning zone is located too far uphill, the clinker will reside for a longer time in the lower-transition zone, being slowly cooled, and thus discharged from the kiln at a temperature of around 1200 C (2192 F).

It is now of interest to discuss how the length of the above-mentioned zones can vary from one kiln system to another.

10.1   LONG WET-PROCESS KILNS

In a wet-process kiln, the feed enters the kiln in the form of a slurry, having a moisture content of approximately 30%. Various types of heat exchangers are used to remove this moisture from the slurry, the one most widely used being chains at the back end of the kiln, shown in Fig. 2.2. Other kiln systems dry the slurry and pelletize the kiln feed before it enters the kiln, using the kiln exit gases for this purpose. Modern, large rotary kilns may use a series of heat exchangers; for example, steel cylinder drying chambers and chains. In some kilns the gases, before they enter the chain section, pass through a section where the interior of the kiln is divided lengthwise into four compartments over a length of approximately 20 ft. This divides the feed bed and thus provides a better exchange of heat between the hot gases and the relatively cold feed. These devices are better known by the term trefoils.

The slurry, when it enters the kiln, has a temperature of approximately 38 C (100 F), rising to a chain discharge temperature of between 200–260 C (392–500 F) in kilns with chain systems that dry all the moisture within the chains. On nodule kilns, i.e., when the feed still contains approximately 8–15% moisture when it leaves the chain system, the feed temperature will be around 100 C (212 F) at this location. Wet-process kilns in which all the free moisture is driven off in the chain system are usually more energy efficient but often show higher kiln exit dust losses, have a tendency toward mud-ring formations in the chains, and produce feed-bed characteristics below the chain zone that are identical to a dry-process kiln. Kilns with chain systems that are designed to form and discharge nodules tend to be easier to control and have less tendency to generate dust flushes (kiln upsets) into the burning zone. From an operator's ease-of-control viewpoint the latter is clearly more desirable. In a typical wet-process kiln, the various zones generally take up the following percentage of total kiln length:

	approximate % of kiln length
Slurry preheat	5%
Evaporation and preheating	40%
Calcination	33%
Burning zone (upper, sintering, and lower-transition zone)	22%

10.2   THE DRY-PROCESS KILN

In the dry process, the feed enters the kiln in dry powder form. As in a wet kiln, the feed passes through a chain section whose primary function is to preheat the feed to a chain exit temperature of around 730 C (1346 F). Thus, whatever free moisture there is in both the feed and the chemically combined water, is driven off within the chain system. Calcination starts a few meters below the chain section. The typical length of various zones in a dry-process kiln are:

	approximate % of total kiln length
Feed drying and preheating	30%
Calcining	47%
Burning zone (upper, sintering, and lower-transition zone)	23%

10.3   THE SEMIDRY (LEPOL) KILN

The drying, the dissociation of the chemical water, the preheating to calcining temperature, and even partial calcination in this system takes place outside the rotary kiln, i.e, in the grate preheater. Hence, the rotary kiln proper has only to complete the calcination and form the clinker compounds. The length of the zones in the Lepol kiln are thus quite different from the aforementioned kilns.

	approximate % of total kiln length
Calcining zone	55%
Burning zone (upper, sintering, and lower-transition zone)	45%

10.4   THE PREHEATER AND PRECALCINER KILNS

These kilns are fed dry powder feed at the upper stage of the preheater cyclone tower. Preheating, dissociation of the chemically combined water, and partial calcination of the feed, here too, takes place outside the rotary kiln. In a preheater kiln, the feed enters the rotary kiln approximately 30–40% calcined whereas in the precalciner kiln this feed can be up to 90% calcined. For this reason, the burning zone in a precalciner kiln is longer and the calcining zone shorter than on a preheater kiln as shown in the following:

	approximate % of total kiln length	
	Preheater	Precalciner
Calcining zone	60%	35%
Burning zone	40%	65%

11.

Coating and Ring Formations
In A Rotary Kiln

11.1 COATING FORMATION IN
THE BURNING ZONE

A good protective coating on the refractory in the burning zone serves to prolong the life of the refractory. Frequent replacement of refractory costs a large amount of money, not only because of the high cost of the refractory, but also because of the loss in production while the kiln is down for lining replacement. Needless to say, frequent renewal of refractory is an undesirable condition in any kiln.

Although refractories in the burning zone have to be replaced from time to time, a kiln operator nevertheless has the capacity to increase (and unfortunately also to decrease) the life of the lining by his ability to control the coating in the burning zone.

Nature of the Coating.

Coating is a mass of clinker or dust particles that adheres to the wall of the kiln, having changed from a liquid or semiliquid to a solidified state. The solidified particles adhere to the surface of the coating (CS in Fig. 11.1), (or the refractory surface (BS) when no coating exists), as long as the temperature of the surface of the coating is below the solidifying temperature of the particles. Coating continues to form until its surface reaches this solidifying temperature. When the kiln operates under such condition at equilibrium, the coating will maintain itself. This means that theoretically no new coating is formed. When this temperature is exceeded, however, the particles on the surface of the coating change again from a solid to a liquid state, and the coating will start to come off.

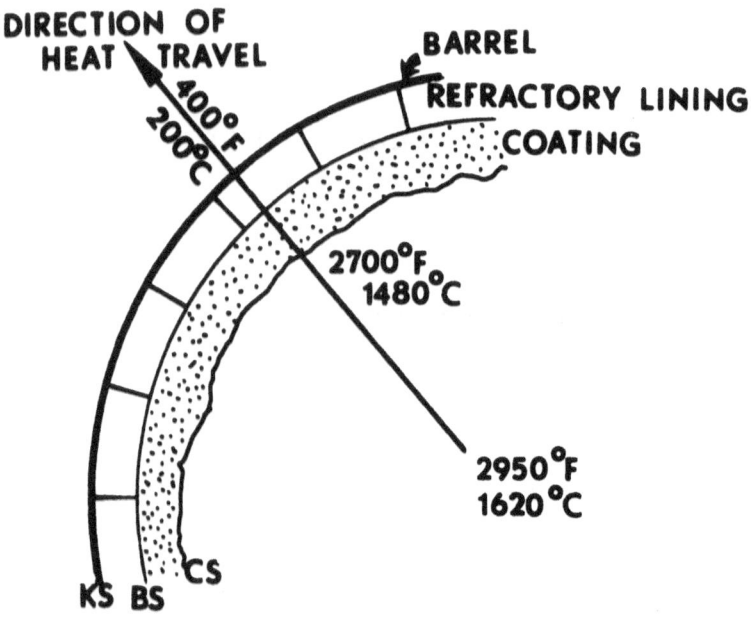

Fig. 11.1   Heat, passing through the shell of the kiln, must be constantly replenished by the flame in order to maintain a condition of equilibrium necessary for coating formation.

There is a temperature drop between the coating surface (CS) and the kiln shell (KS), the heat flowing in the direction indicated by the arrow in Fig. 11.1. (Heat always travels from a place or body of high temperature to a place of body of lower temperature.) This heat transfer is governed to a great extent by the conductivity of the refractory and the coating. The better the conductivity of the refractory, the better the chance of coating formation, explained by the fact that the more heat that travels in the direction of the arrow, the lower the temperature will be at the surface of the coating. Because the coating consists of particles that have changed from a liquid to a solid state, the amount that any kiln feed liquefies at clinkering temperature plays a very important role in coating formation. A kiln feed with a high liquid content at clinkering temperatures is more effective in coating formation than a feed low in liquid. Kiln feeds with a high liquid phase (easy-burning mixes) have a high content of fluxes: the iron, alumina, magnesia, and alkalies. On the other hand, hard-burning mixes (low in iron, alumina, magnesia, and alkalies, and high in silica and lime) do not have a favorable influence on coating formation. Alkalies entrained in the gas stream promote the formation of coating (unfortunately rings also) because of their high fluxing characteristics.

Because the surface temperature is probably the most important factor in the formation of a coating, it is obvious that the flame itself has a significant effect on coating formation because the shape of the flame directly governs the surface temperature at any given point in the burning zone. A flame that is too short, snappy, and wide can erode the coating because of the great heat released over a short area with this kind of flame. A long flame is more favorable to coating formation in the burning zone. In Chapter 6 it was pointed out that short flames are desirable for better control of the burning operation, but the flame should be shortened only to the extent that it will not harm the coating.

Once all the factors have been taken care of and the foundation for a good protective coating established, it is then up to the kiln operator to control the coating during operation of the kiln. It is his responsibility to form and maintain a good solid coating in the burning zone.

Operating Conditions.

Operating conditions are just as vital for coating formation as all the other factors mentioned above. Assume that a kiln will be operated from one extreme of temperature to the other, that is, a cold, a normal, and a badly overheated kiln; that the same kiln-feed composition is burned in all three examples; that the solidifying temperature is 2400 F (1315 C); and that 24% liquid is formed at the point of investigation, under ideal operating condition.

First, consider the cold kiln (Fig. 11.2A). In this case almost no coating is formed. The coating surface temperature as well as the feed temperature is too low to produce the necessary amount of liquid matter that would promote coating formation. The condition in this example is commonly referred to by kiln operators as the kiln being in a "hole." This example also supports the widely known fact that no new coating can be formed while the kiln is cold.

In the normal kiln (Fig. 11.2B), enough liquid (24%) is present to form a coating. Temperature of the coating when it emerges from the feed bed, as well as when in contact with the feed, is below the solidifying temperature of the feed particles. The particles will adhere to the wall and solidify, and will continue to do so as long as the surface temperature of the coating remains below the solidifying temperature of 2400 F (1315 C). Whenever the wall reaches this temperature no new coating will form. The coating is in equilibrium.

In the hot kiln (Fig. 11.2C), because of the extremely high temperatures of the feed and the coating, too much liquid is formed. As all temperatures are above the solidifying temperature, the coating transforms from a solid back to a liquid again. In such a condition, coating will come off, and the feed because of its high liquid content will "ball up." Needless to say, this condition is extremely harmful to the kiln and to the refractory.

Most basic refractories, and especially the dolomite liners, are not able to withstand prolonged exposure to the high flame temperatures without this protective coating. As was mentioned in the previous chapter, the burning zone is divided into three subzones namely the upper-transition, the sintering, and the lower-transition zones. Because of the lower liquid content in the feed and because of the frequent temperature changes, the upper- and lower-transition zones are areas where formation and maintenance of coating is the most unstable. Shifting burning zone locations produce a similar shift in the location where coating is formed; thus, unstable coating conditions are most frequently observed in the upper and lower end of the burning zone. This is clearly supported by the fact that most rotary kilns experience the most frequent refractory failures in these two critical areas. It should be noted that since the upper and lower burning zones are also within the vicinity of the first and second tires, brick failures are not only the result of variations in burning-zone conditions but, are also often the direct result of excessive tire clearance and shell ovality. Both the frequent falling out of coatings in these areas and the formation of too much coating can lead to troublesome ring formations.

Ring formations in the lower-transition zone (i.e., at the kiln discharge) are referred to as nose rings. Others refer to these as ash rings when the kiln is coal fired. Ring formations in the upper-transition zone are referred to as clinker rings. These ring formations can in many instances be so severe that they force operators to shut down the kiln and shoot these rings

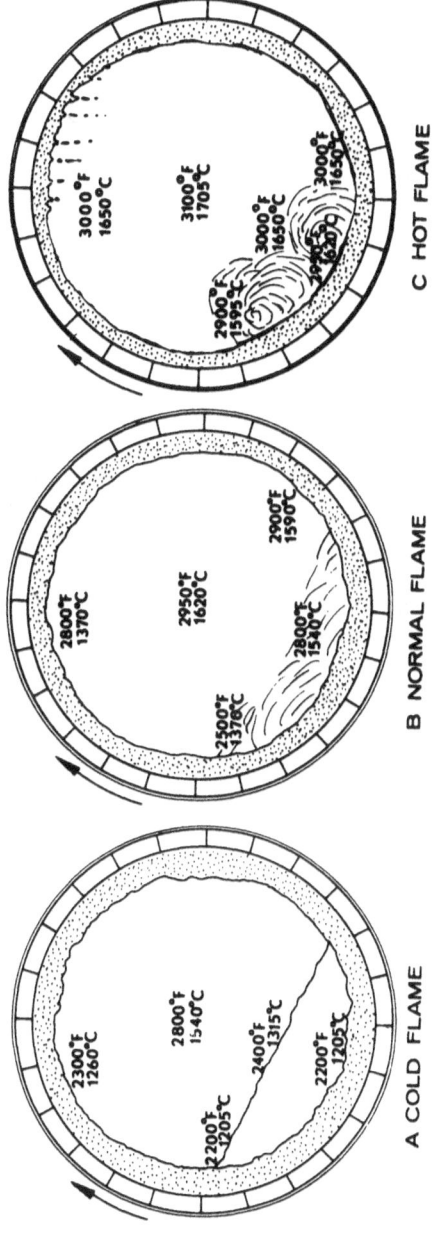

Fig. 11.2   Kiln A, being in a hole, is not hot enough for proper formation of coating; good coating is formed in kiln B at normal operating temperatures; the high temperature in kiln C causes too much liquid to form and can result in a serious loss of coating.

out with an industrial gun. The Cardox system has been successfully used for many years in Europe on several kilns to remove such rings. These devices are affixed to the kiln shell in strategic locations and use CO_2 cartridges to blast the rings while the kiln has only to be stopped for a short interval to load and trigger the cartridges.

Much research work has been done on the probable causes of these ring formations in the burning zone. The possible causes are many and no one single factor has yet been found that would be the main cause for all the rings formed. What seems to be true for one particular kiln might be competely wrong for another kiln. This is clearly explained in the following example: On many coal-fired kilns, operators have found a relationship between the fusion temperature of the coal ash and the frequency of ring formation. There appears to be more ring formation when the fusion temperature is low, i.e., when the ash contains larger amounts of fluxing iron and alumina and less silica. However, this could not be the only cause for such ring formations because natural gas- and oil-fired kilns, which have no ash deposits in the burning zone, can have just as many ring problems as the coal-fired kilns. Hence, solutions for the elimination of rings in the burning zone are predominantly found by a process of elimination. First, all probable causes are listed and then each suspected cause is eliminated or changed until hopefully an answer is found. From personal experience, the author has found the following factors to be possible contributors to ring formations in the burning zone:

POSSIBLE CAUSES FOR RING FORMATION IN THE BURNING ZONE

a) Coal fineness too coarse.
b) Low fusion temperature of coal ash.
c) Kiln feed high on liquid content (silica, A/F ratios as well as lime saturation factor low).
d) Incomplete calcination of the feed as it enters the burning zone (frequent dust flushes into and poor calcining conditions behind the burning zone).
e) Frequent changes in chemical composition and fineness of kiln feed.
f) Excessive dust generation in the cooler and burning zone (including changes in dust insufflation rates on wet-process kilns).

g) Kiln speed too slow and feed loading too high in normal operation.

h) Excessive variations of flame temperature and length during normal operation.

i) Frequent changes in secondary air temperatures.

j) Excessive frequency of kiln-operating upsets (burning zone temperature and location varies too frequently and by too large a range).

k) Increased volatility of, and frequent changes in, alkali and sulfur contents in the fuel and feed.

Others have found other reasons for ring formation in the burning zone. Thus, the list could possibly be expanded to over 30.

It is of interest that half of the cited factors can be somehow controlled by the kiln operator and action taken to stabilize the flame and the kiln operation that might be beneficial in lessening the frequency of ring formation.

11.2   COATING AND RING FORMATION UPHILL OF THE BURNING ZONE

Less frequent but nevertheless equally troublesome are the so-called feed rings that form in the calcining zone of the rotary kiln. Wet-process kilns often experience so-called mudring formations in the chain section too. Finally, many preheater kilns and Lepol kilns experience ring formations and build-up problems at the feed inlet and in the lower preheat cyclone stage.

In investigations on this subject it has been found that the majority of these rings and heavy coatings in dry- and wet-process kilns are associated with one of the following factors:

a) Internal cycle of the volatile constituents from the kiln feed and fuel (alkalies, sulfur, chlorides).

b) Kiln-feed fineness.

c) Irregular and insufficient control (frequent fluctuations) of the feed-end temperature and kiln draft.

d) Excessive dust generation within the rotary kiln proper.

Analysis of the materials from these rings or excessive coating buildup invariably showed high contents of calcium sulfates, potassium chlorides, and/or alkali sulfates. Efforts to alter the internal and external cycle of volatile components in the gas or feed stream have in many instances resulted in less frequent ring formations. Although the aforementioned reasons apply to dry- and wet-process kilns, there have been many reports by others that have found a similar relationship in preheater and Lepol kilns.

In one case of a wet-process kiln, it was found that mudring formation was caused by a large percentage of the coarse fraction in the kiln feed which contained predominantly free silica. In another case, mudrings were initiated by the kiln operator who did not excercise sufficiently tight control over the feed-end temperature, i.e, he allowed the temperature to vary within a large range. Dust insufflation of the introduction of dust through scoop feeders below the chain section can also be a primary cause for mudring formations in wet-process kilns. Last but not least, mudrings were also allowed to be formed simply by having the wrong type of chain-system design or chain-link sizes.

References

1. Waddell, Joseph J. 1962. *Practical Quality Control for Concrete.* New York:McGraw-Hill Book Co.
2. Bogue, R.H. 1929. *Industrial Engineering Chemistry*, Anal. Ed. 1:192.
3. Lea, F.M. and Parker, T.W. 1935. *Building Research*, Technical Paper 16, London.
4. Kuehl, H.1929. *Zement*, 18, 833.
5. Lea, F.M. and Desch, C.H. 1935. *The Chemistry of Cement and Concrete.* Longmans, Green & Co.

12.

The Air Circuit in a Rotary Kiln

For a kiln operator, three variables that are related to air and gas flows anywhere within the kiln system are of special interest. These are:

a) Volume and velocity
b) Temperature
c) Pressure

To determine the *volume* of a gas in a duct or flue, it is necessary to measure both the temperature and the velocity. *Velocity* is determined by means of a Pitot tube and differential pressure gauge. Other instruments used for this purpose are: *anemometers* (when the velocities are less than 3 m/s), *orifice disk* or *Venturi meters* (when gases move at high velocity or pressure through small pipes), or *S-tubes* that perform the same function as Pitot tubes except that they are used for measurements of dust-entrained gas streams. Strict rules apply and must be followed in order to obtain meaningful and accurate results when velocity measurements are made in a duct.

Pressure is measured and expressed by three different methods: *static pressure* is measured at a right angle to the gas stream; *impact pressure* (often referred to as total pressure) is measured by pointing the tube directly into the center of the gas flow; and finally, the *velocity pressure* (often referred to as dynamic pressure) is measured by subtracting the static pressure from the impact pressure, as shown in Eq. (12-1).

$$P_v = P_I - P_S \qquad (12\text{-}1)$$

155

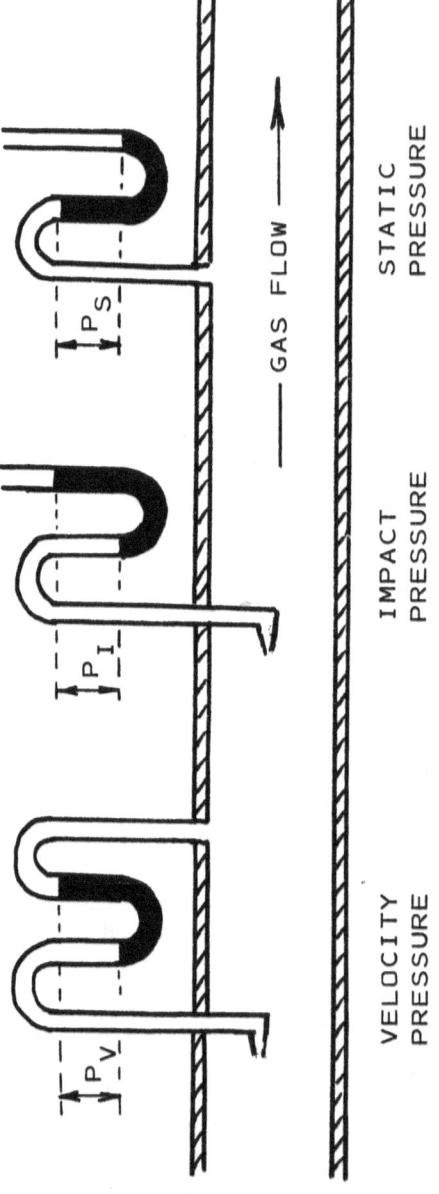

Fig. 12.1 Pressure measurements.

Instruments used for measuring pressures are: dial gauges, U-tubes, and inclined draft gauges—the latter being the most sensitive and accurate of the three.

After accurate pressure and temperature measurements have been taken one can calculate the air or gas velocity by the following formula:

$$V_S = 2gP_V \qquad\qquad (12\text{-}2)$$

where:

V_S = gas velocity per second
g = acceleration of gravity
P_V = velocity pressure

Either English or metric units can be used in the above formula with the appropriate use of the English or metric system gravitational constant.

The volume of gas passing through the duct is calculated by the formula:

$$Q = 60AV_S \qquad\qquad (12\text{-}3)$$

where:

Q = volume of gas flow per minute
V_S = average gas velocity (per second)
A = effective cross-sectional area of the duct

Air movement through the kiln can be considered, for all practical purposes, to take place uniformly from introduction of the air in the cooler to final discharge at the stack. In accordance with the gas laws, however, the physical state of the air undergoes changes in temperature, volume, and pressure while traversing the kiln. The reasons for these changes is that the chemical composition of the air itself changes in the process, and the spaces through which the air travels are not uniform throughout the system. Figs. 12.2–12.4 show schematic diagrams of complete air circuits of the various kiln systems in use in the cement industry.

For convenience, it is customary to divide this air flow into three successive circuits in the following order:

a) The circuit for cooling of the clinker and introduction of combustion air into the kiln. This is the cooler circuit.
b) The circuit in which combustion, calcination, and drying takes place. This is the kiln circuit.

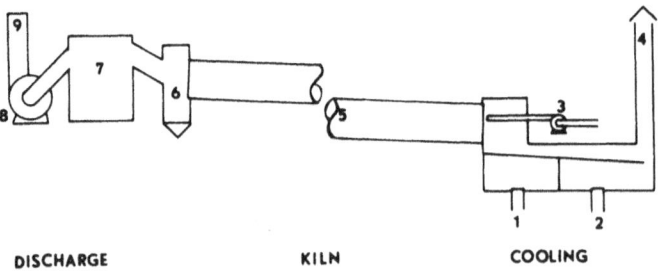

DISCHARGE   KILN   COOLING

Fig. 12.2   Air circuit on wet- and dry-process kilns—1 and 2 are air inlets into the undergrate chamber of the clinker cooler; 3 is primary air fan; 4, excess cooler-air stack; 5, kiln; 6 and 7, dust collectors; 8, induced draft fan; 9, stack.

   c) The circuit in which the air and gases are released from the kiln and pass through the heat recovery and dust-collecting units. This is the discharge circuit.

In considering the air circuit in the kiln, it is necessary to look at all three divisions as a whole because one leads naturally to the next. The whole concept of air flow is one of a complete unit, as any adjustment or change in one circuit will effect a change in the entire air circuit. Another important factor is that it is almost impossible to avoid the entrance of so-called cold parasite air into the kiln system. This air enters through badly closed openings around the kiln hood and around the back-end ducts. Not only does this cut down the efficiency of the induced draft fan, but the cold air entering through the hood and nose area reduces the fuel efficiency. Drastic changes in the air flow take place when doors anywhere in the system are left open for a prolonged length of time. Because this could lead to upset kiln conditions, it is important to advise the operator in advance when such an action is to be undertaken.

12.1   THE COOLER AIR CIRCUIT

The clinker cooler performs two functions: it must cool the hot clinker discharged from the kiln and supply the kiln with the necessary air for

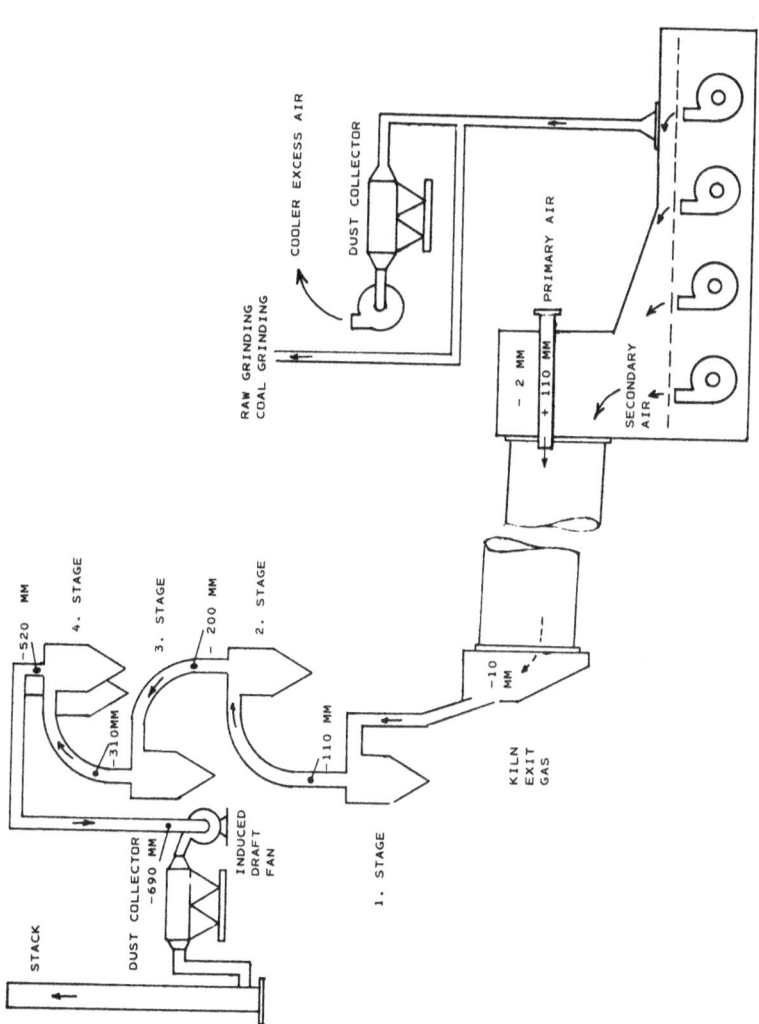

Fig. 12.3 Air circuit in suspension preheater kiln.

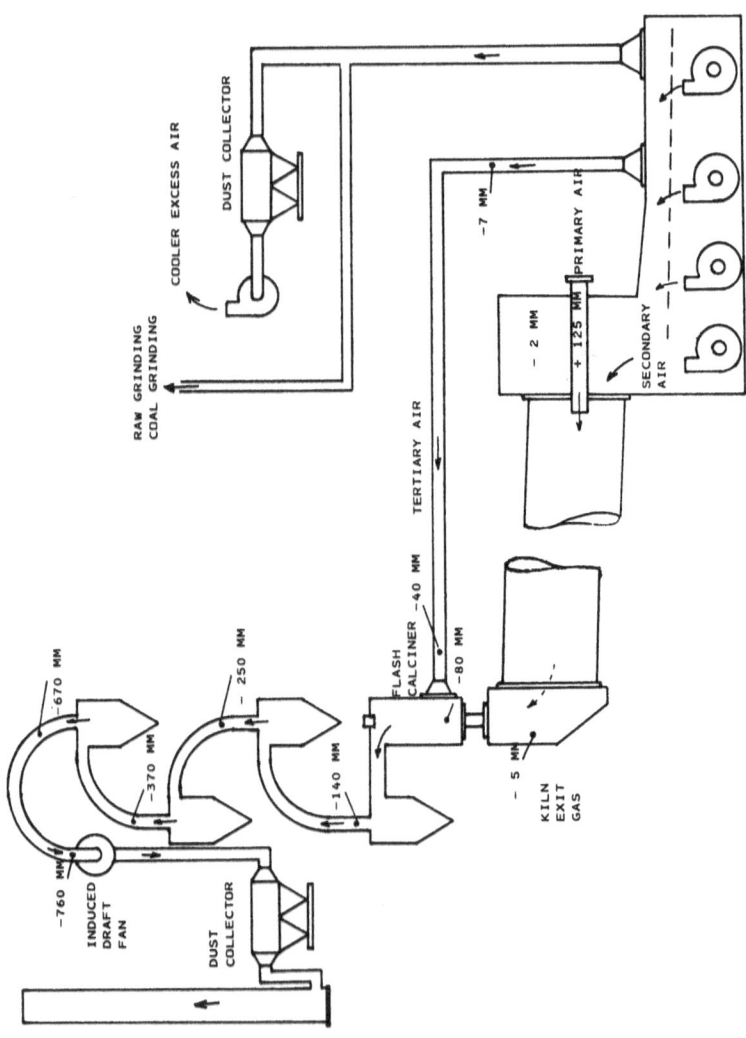

Fig. 12.4 Air circuit in precalciner kiln.

combustion. In doing so, valuable heat from the clinker is recuperated and enters the kiln as hot secondary air. The most common clinker coolers in use are:

a) inclined reciprocating grate coolers
b) planetary coolers
c) continuous traveling grate coolers
d) rotary coolers

On kilns with large clinker outputs or special cooling problems, additional cooling of the clinker is often required after it discharges from these coolers. This is being done by adding a so-called after-cooler to the above units, the best known type of this kind being the G-cooler manufactured by Claudius-Peters. Such after-coolers are also viable alternatives (instead of installing completely new coolers) in cases when a kiln is being converted from wet to dry, dry to preheater, or from preheater to precalciner status, all of which produce a need for increased cooler capacity due to anticipated higher clinker output rates. Unlike conventional coolers, there is no direct contact between the clinker and the air in a G-cooler, hence, this type of equipment is almost free of dust emissions. G-coolers provide the final clinker cooling, receiving clinker at a temperature of around 350 C (660 F) and discharging clinker at a temperature of approximately 100 C (212 F).

The inclined *reciprocating grate cooler* is the most widely used type of cooler today. This type of cooler derives its name from the reciprocating (back-and-forth) movement of the rows of cooler grates that push the clinker toward the discharge end. As many operators can attest to, this type of cooler can not properly and efficiently perform its function when the clinker is very fine as in the case where the kiln is in an upset condition. Fig. 12.5 shows a schematic cross section of a *closed-loop traveling grate cooler*. This type of reciprocating grate cooler has found some applications in wet-process kilns where combustion air requirements in the kiln are usually very high. Circulating warm air from the lower section of the cooler and using it for cooling air in the upper compartment lowers the total excess air that has to be vented to the atmosphere and is thus beneficial toward controlling the emissions from the cooler. However, this system has its application limitations and is usually not suitable for large dry and preheater kilns that are energy efficient and produce a large amount of clinker per hour. Nevertheless, it is known that there is at least one precalciner kiln that uses this kind of a system successfully in the United

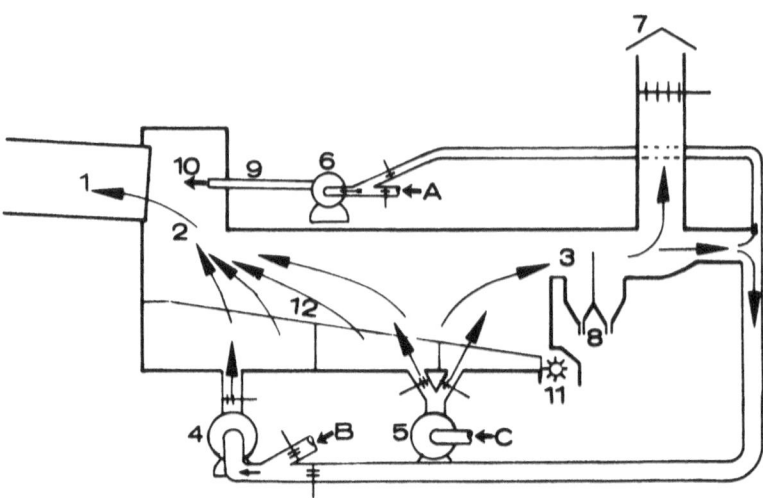

Fig. 12.5 Cross section of closed loop traveling grate cooler, showing: 1, kiln; 2, secondary air; 3, circulatory air; 4, hot air fan; 5, cold air fan; 6, primary air fan; 7, excess cooler-air stack; 8, dust collectors; 9, burner pipe; 10, primary air; 11, clinker crusher; 12, traveling grates; A, B, and C are ambient air inlets.

States. There is a limit to which air can be recycled within the cooler compartments in order to maintain rapid quenching in the first cooler compartment. Coolers equipped with such a system often experience rapid cooler fan blade wear when the recycled air contains a large amount of dust.

Cooler operating problems have become more pronounced as kilns started to get larger and their output rates increased. Most grate coolers require the use of a much larger amount of air for cooling the clinker than what is needed for combustion in the kiln. Hence, modern coolers operate with larger cooler excess-air volumes that are not usable for combustion in the kiln. To prevent these heat losses, this excess air is usually used for drying the kiln feed and/or coal and, in precalciner kilns, is transferred to the flash furnace to be used as combustion air in the auxiliary firing unit.

Planetary coolers are cooling cylinders that are attached to the circumference of the rotary kiln. There are no moving parts on these types of coolers, no fans to control, and all the air that passes through these tubes enters the kiln and is used for combustion. It's the simplest system for cooling clinker and there are no control functions except the rotational speed of the kiln itself. Planetary coolers are equipped with lifters, agitators, and blades to promote the heat exchange between the hot clinker and the cold air. With all the air entering the kiln, there are no dust-emission problems from these types of coolers. On larger units, noise pollution can, however, become a problem from the clinker rattling in the tubes. Planetary coolers require a larger I.D. fan than kilns of equal size equipped with grate coolers but the total power requirements are much lower when one considers all the air fans usually found on grate coolers. The disadvantage of planetary coolers is that most will discharge the clinker at a higher temperature than compatible grate coolers.

The *continuous traveling grate coolers* are similar to the reciprocating grate coolers except that they use a continuous moving grate belt. Advantages of this type of cooler are that individual worn or damaged plates can be quickly replaced without needing to stop the kiln. Also, since the belt travels from the hot to the colder part of the cooler and returns underneath where it is being cooled by the cold air, overheating is less likely to occur here than in reciprocating grates.

Rotary coolers were first used over 80 years ago, however, in order to be efficient in today's modern kilns they would have to be of such size as to make them financially unjustifiable. There is some doubt that this type of a cooler would ever make a comeback in the cement industry despite its simplicity of operation.

Every cooling system must have an air balance to ensure proper cooling of the clinker as well as proper combustion in the kiln. Any excess air from the cooler not needed for combustion has to be vented to the atmosphere or must be used for other thermal work.

Efficient operation of a cooler requires that the minimum amount of air with the lowest possible temperature be vented to the atmosphere while at the same time the clinker is cooled to the temperature that ensures trouble-free transport of the clinker to the storage area after it leaves the cooler. Dampers and regulators must be so adjusted that no overheating of the cooler equipment can take place, and the primary air and secondary air are blended in the proper proportions when they enter the kiln. Temperature of the secondary air entering the kiln should be as high as possible in order to recoup a maximum amount of heat from the clinker, provided the cooler and kiln nose are not permitted to become overheated.

On grate coolers, cooling is accomplished by forcing ambient air upward through the grates and the bed of material. To obtain optimum cooling it is important that the proper bed resistance is maintained. This insures proper retention time for the clinker and efficient heat transfer. As the cooling air passes through the clinker bed and acquires a higher temperature, its volume is expanded. As the air expands, more force is required to push this air through the bed. Since the material is hottest at the inlet to the cooler, it follows that the cooling air expands the most here and requires the highest pressure at this inlet side. Further down, the heat-transfer rate becomes less, causing less expansion of the air and thus requiring less pressure to get through the bed. This is the reason for installing several compartments in these coolers and equipping each with a fan of diminishing static-pressure rating. When this important factor is overlooked in the initial design phase, or when the cooler is not properly and efficiently compartmentalized, cooling efficiency of the unit as a whole is sacrificed.

After the cooling air passes through the bed of clinker, the air needed for combustion is drawn into the kiln as the so-called secondary air by the I.D. fan or forced into the kiln as primary air by the primary air fan. The remaining air is either vented to the atmosphere or used for drying coal or raw feed. In the precalciner kiln, this excess air is transferred to the flash furnace by means of the tertiary air duct.

It is a fundamental requirement that the air forced into the cooler is applied to these areas where it can do the most useful thermal work. In thermodynamics, the best heat-transfer potentials take place where the temperature differential is the greatest. It follows that it is necessary to apply the maximum amount of air in the upper (hot) compartments and a minimum amount in the lower (cold) compartments. This can be clearly demonstrated by using the common heat-transfer Eq. (12-4):

$$\text{Air} \qquad\qquad\qquad \text{Clinker}$$

$$(1)\ (\text{sp.ht.})\ (t_2 - t_1)\ =\ (1)\ (\text{sp.ht.})\ (T_1 - T_2) \qquad\qquad (12\text{-}4)$$

Inserting the appropriate values for the variables, this demonstrates that one pound of ambient air has the capacity to lower the clinker temperature in the first compartment from 1260 C (2300 F) down to 748 C (1378 F), i.e., producing a temperature drop of 512 C (922 F) in the clinker. However, the same pound of ambient air, applied to the last cooler compartment will be able to drop the temperature of a "colder" clinker of, e.g., 300 C (572 F) by only 13 C (23 F). In other words, the heat transfer in the first compartment is 30 to 40 times more efficient than the heat transfer in the lower compartment. This serves as a clear reminder to operators that they should always first make sure that the air is properly applied in that part of the cooler where it can do the most thermal work. This becomes especially important in situations where a cooler exhibits very high excess-air volumes that must be vented to the atmosphere. The same reasoning must be applied when the clinker cooler consistently discharges insufficiently cooled clinker that produces high finish-grinding temperatures and could lead to quality-control problems in the cement. Many cooler-operating problems have been, and can continue to be, solved by applying this important rule of cooler operation. Damage to cooler components often occurs when an operator neglects to recognize when the air flow to individual cooler compartments is not properly adjusted to prevailing operating conditions.

It is a recommended procedure that the plant engineering staff checks the cooler air flows and establishes a cooler air balance at least once per year to keep up-to-date information regarding the cooler's efficiency.

12.2   THE ROTARY-KILN AIR CIRCUIT

The amount of air drawn through the rotary kiln is governed and controlled *solely* by the induced draft fan (I.D. fan) located at the feed end of the kiln. The origin of the air going to the kiln is:

1) the hot secondary air emanating directly from the cooler,
2) the moderately heated primary air entering through the burner pipe, and
3) the parasite air that enters through leaks and openings in the kiln-hood area.

The ultimate objectives are: a) to induce a maximum amount of hot secondary air, b) to use only the minimum, safe amount of primary air that is needed for fuel conveyance and flame control, and finally c) to remove or reduce the presence of parasite air. For every pound of cold air (e.g., 30 C) drawn through these unnecessary in-leakage points in the hood, there is a corresponding pound of valuable hot secondary air of around 800 C that will not be put to useful work in the kiln. The end-result is that a higher volume of excess cooler air has to be vented into the atmosphere. The temperature of this air is also being increased resulting in the waste of more valuable heat. The kiln operator must always try to stabilize the air flow through the kiln. Unless the kiln draft is controlled within reasonable limits, a uniform firing rate and stable burning conditions can never be obtained.

A certain definite amount of air is required for complete combustion in the kiln, which means that close control over the fuel-to-air ratio has to be exercised. Incomplete combustion, caused by the admission of too little air, results in the loss of some unburned carbon because a portion of the carbon in the fuel is burned to carbon monoxide (CO) instead of carbon dioxide (CO_2). To make sure that this does not take place a small amount of excess air (approximately 5%) is introduced into the kiln. On the other hand, too large an amount of excess air could lead to serious heat losses because this air, not needed for combustion, raises the temperature of the gases leaving the kiln at the rear. This rise in temperature in turn affects the heat profile in the kiln, because of the change it causes in the drying, calcination, and burning-zone conditions. Once a kiln operation is stabilized, the following three basic rules must be observed in order to maintain the kiln in such a condition:

a) Complete combustion conditions must prevail at all times while a kiln is in operation, thus fuel and air rate changes can be undertaken only to the extent that no formation of carbon monoxide will result after the rates have been changed.

b) The maximum amount of excess air present in the kiln should not exceed the amount that will cause heat losses to the rear to surpass the predetermined limit.

c) Temperatures of the air entering the kiln and leaving the kiln should be held as nearly constant as possible in order not to upset the heat profile in the kiln.

By weight, each unit of fuel burned (lb or kg) requires approximately 10.5 units (lb or kg) of combustion air. Air, at standard conditions, has a specific volume of approximately 0.7735 m³/kg (12.39 ft³/lb). Hence, to burn 1 kg of fuel requires approximately 8.12 m³ of air for combustion in the kiln. Likewise, burning 1 lb of fuel requires approximately 130 ft³ of air at standard conditions. The gases produced during the combustion process travel toward the rear of the kiln and are enriched with:

a) carbon dioxide evolved from the feed during calcination,

b) water from the evaporation of the moisture in the feed,

c) volatile constituents that have evolved (e.g., sulfur and alkalies), and

d) dust evolved within the system.

Clearly it stands to reason that a wet-process kiln, where all the calcination (CO_2) and all the evaporation of the moisture in the fed takes place within the rotary kiln proper, has the highest gas volumes per unit weight of clinker produced. Conversely, the precalciner kiln, where most of the calcination is done in the preheater tower and no moisture is present in the feed, has the lowest volume of kiln gases leaving the rotating kiln.

Gas-flow rates can be either calculated based on theoretical parameters or actually measured by the aforementioned measuring methods. However, actual measurements, particularly kiln exit-gas flows, will always be suspect for accuracy unless strict adherence to established measuring rules is practiced. It is also very difficult to establish the actual cross-sectional area of the duct, at the point of measurement, because there usually are no accurate means to determine the size of the coating buildup in the duct. It is therefore good practice for an engineer to calculate the theoretical flow rates as well as to ascertain that the actual measured values correspond closely to the theoretical results.

In the following tables (Tables 12.1–12.4), kiln-air and gas-flow rates of the four main kiln systems discussed in this book are shown. Operating condition (A) in Tables 12.1 and 12.2 refers to what is considered efficient operations. Operating condition (B) in Tables 12.3 and 12.4 has base data altered to show how the gas weights and volumes can change on the same kilns as a result of less efficient operating conditions. These tables have been developed on theoretical considerations only for the purpose of demonstrating how changes in fuel and calcining rates can change the gas-flow rates in the kiln.

Each individual type of kiln design has an established, by design, pressure drop across the system. Fans are thus designed to overcome this pressure drop or resistance under normal operating conditons. Unfortunately, these pressure drops do not remain constant since added restrictions in the form of buildups and rings can significantly increase this pressure drop. Pressure-sensing devices that monitor pressure changes within the system are a useful tool for operators to detect formation of such restrictions. This is discussed in detail in the chapter on kiln control.

TABLE 12.1

WEIGHTS OF KILN AIR AND GAS FLOWS (typical efficient operation)

Weights of Kiln Air and Gas-Flow Rates

Base Data	Operating Condition (A)							
	Wet-process kiln		Dry-process kiln		Preheater kiln		Precalciner kiln	
	English	Metric	English	Metric	English	Metric	English	Metric
Fuel rate (lb/sh.t or kg/t)	396	198	332	166	238.4	119.2	244	122
Specific fuel consumption (MBtu/t or kJ/kg)	4.95	5757	4.15	4827	2.98	3466	3.05	3547
Calcination done in kiln (%)	1	1	1	1	0.6	0.6	0.2	0.2
Fuel fired in flash furnace (%)	0	0	0	0	0.6	0.6	0.5	0.5
(values are in terms of lb/sh.t clinker or kg/t clinker)								
Total air required for combustion in kiln	4158	2079	3486	1743	2503	1252	1281	641
Secondary air	3326	1663	2789	1394	2253	1126	1153	576
Primary air	665	333	558	279	200	100	102	51
Infiltrated air at kiln hood	166	83	139	70	50	25	26	13
Combustion product in kiln	4633	2317	3884	1942	2789	1395	1427	714
Gases from the feed	2732	1366	1196	598	643	322	214	107
Total gases leaving rotary kiln	7365	3683	5080	2540	3433	1716	1642	821
Infiltrated air at kiln back end	1395	698	1176	588	624	312	344	172
Total air required for combustion, flash furnace	0	0	0	0	0	0	1281	641
Combustion product in preheater tower	0	0	0	0	0	0	1427	714
Gases from the feed (preheater tower)	0	0	0	0	519	260	948	474
Total gases leaving the kiln system (i.e., going to I.D. fan)	8760	4380	6256	3128	4576	2288	5642	2821

TABLE 12.2

VOLUMES OF KILN AIR AND GAS FLOWS (typical efficient operation)

Volumes of Kiln Air and Gas-Flow Rates

Base Data	Operating Condition (A)							
	Wet-process kiln		Dry-process kiln		Preheater kiln		Precalciner kiln	
	English	Metric	English	Metric	English	Metric	English	Metric
Fuel rate (lb/sh.t or kg/t)	396	198	332	166	238.4	119.2	244	122
Specific fuel consumption (MBtu/t or kJ/kg)	4.95	5757	4.15	4827	2.98	3466	3.05	3547
Calcination done in kiln (%)	1	1	1	1	0.6	0.6	0.2	0.2
Fuel fired in flash furnace (%)	0	0	0	0	0	0	0.5	0.5
(in terms of SCF/sh.t and Nm³/kg cl, standard condition)								
Total air required for combustion in kiln	51,518	1.61	43,192	1.35	31,015	0.97	15,872	0.50
Secondary air	41,214	1.29	34,553	1.08	27,913	0.87	14,284	0.45
Primary air	8,243	0.26	6,911	0.22	2,481	0.08	1,270	0.04
Infiltrated air at kiln hood	2,061	0.06	1,728	0.05	620	0.02	317	0.01
Combustion product in kiln	59,212	1.85	49,643	1.55	35,647	1.11	18,242	0.57
Gases from the feed	32,210	1.01	12,881	0.40	6,775	0.21	2,215	0.07
Total gases leaving rotary kiln	91,423	2.85	62,524	1.95	42,422	1.32	20,458	0.64
Infiltrated air at kiln back end	17,284	0.54	14,571	0.45	7,731	0.24	4,262	0.13
Total air required for combustion, flash furnace	0	0.00	0	0.00	0	0.00	15,603	0.49
Combustion product in flash furnace	0	0.00	0	0.00	0	0.00	18,242	0.57
Gases from the feed (preheater tower)	0	0.00	0	0.00	5,336	0.17	9,793	0.31
Total gases leaving the kiln system (i.e., going to I.D. fan)	108,707	3.39	77,094	2.41	55,489	1.73	68,357	2.13

TABLE 12.3

WEIGHTS OF KILN AIR AND GAS FLOWS (assumed inefficient operation)

Weights of Kiln Air and Gas-Flow Rates

Base Data	Operating Condition (B)							
	Wet-process kiln		Dry-process kiln		Preheater kiln		Precalciner kiln	
	English	Metric	English	Metric	English	Metric	English	Metric
Fuel rate (lb/sh.t or kg/t)	424	212	355.2	177.6	255.2	127.6	260.8	130.4
Specific fuel consumption (MBtu/t or kJ/kg)	5.3	6164	4.44	5164	3.19	3710	3.26	3792
Calcination done in kiln (%)	1	1	1	1	0.85	0.85	0.35	0.35
Fuel fired in flash furnace (%)	0	0	0	0	0	0	0.4	0.4
(values are in terms of lb/sh.t clinker or kg/t clinker)								
Total air required for combustion in kiln	4452	2226	3730	1865	2680	1340	1643	822
Secondary air	3562	1781	2984	1492	2412	1206	1479	739
Primary air	712	356	597	298	214	107	131	66
Infiltrated air at kiln hood	178	89	149	75	54	27	33	16
Combustion product in kiln	4961	2480	4156	2078	2986	1493	1831	915
Gases from the feed	2732	1366	1196	598	911	456	375	188
Total gases leaving rotary kiln	7693	3846	5352	2676	3897	1949	2206	1103
Infiltrated air at kiln back end	1550	775	1470	735	780	390	430	215
Total air required for combustion, flash furnace	0	0	0	0	0	0	1095	548
Combustion product in preheater tower	0	0	0	0	0	0	1221	610
Gases from the feed (preheater tower)	0	0	0	0	251	125	787	394
Total gases leaving the kiln system (i.e., going to I.D. fan)	9243	4621	6822	3411	4928	2464	5739	2870

TABLE 12.4

VOLUMES OF KILN AIR AND GAS FLOWS (assumed inefficient operation)

Volumes of Kiln Air and Gas-Flow Rates

Base Data	Operating Condition (B)							
	Wet-process kiln		Dry-process kiln		Preheater kiln		Precalciner kiln	
	English	Metric	English	Metric	English	Metric	English	Metric
Fuel rate (lb/sh.t or kg/t)	424	212	355.2	177.6	255.2	127.6	260.8	130.4
Specific fuel consumption (MBtu/t or kJ/kg)	5.3	6164	4.44	5164	3.19	3710	3.26	3792
Calcination done in kiln (%)	1	1	1	1	0.85	0.7	0.35	0.2
Fuel fired in flash furnace (%)	0	0	0	0	0	0	0.4	0.5
(in terms of SCF/sh.t and Nm^3/kg cl, std. condition)								
Total air required for combustion in kiln	55,160	1.72	46,210	1.44	33,200	1.04	20,357	0.64
Secondary air	44,128	1.38	36,968	1.15	29,880	0.93	18,322	0.57
Primary air	8,826	0.28	7,394	0.23	2,656	0.08	1,629	0.05
Infiltrated air at kiln hood	2,206	0.07	1,848	0.06	664	0.02	407	0.01
Combustion product in kiln	63,399	1.98	53,112	1.66	38,159	1.19	23,398	0.73
Gases from the feed	32,210	1.01	12,881	0.40	9,598	0.30	3,877	0.12
Total gases leaving rotary kiln	95,609	2.98	65,993	2.06	47,757	1.49	27,275	0.85
Infiltrated air at kiln back end	19,205	0.60	18,213	0.57	9,664	0.30	5,328	0.17
Total air required for combustion, flash furnace	0	0.00	0	0.00	0	0.00	13,341	0.42
Combustion product in flash furnace	0	0.00	0	0.00	0	0.00	15,599	0.49
Gases from the feed (preheater tower)	0	0.00	0	0.00	2,580	0.08	8,131	0.25
gases leaving the kiln system (i.e., going to I.D. fan)	114,814	3.58	84,206	2.63	60,001	1.87	69,674	2.17

12.3   THE DISCHARGE AIR CIRCUIT

Once the gases leave the kiln system (preheater tower), there remains the task of removing the dust in these gases before they can be vented to the atmosphere. Multiclones, cyclones, electrostatic precipitators, and/or dust-bag houses are installed for this purpose. Without entering into a detailed description of various dust-collecting systems, it can generally be said that all these units cause an appreciable loss of pressure in the gas stream, which makes a large capacity of the induced draft fan a necessity. The more dust-collecting units and the shorter the chimney or stack, the more energy and fan capacity required to pull the gases through the kiln system.

The operator should keep in mind that the I.D. fan is designed to handle the hot kiln gases given off during normal operating conditions. If the fan is called upon to move an equal volume of relatively cold gas, it could be overloaded because the cold gas has a higher density (is heavier) than the hot gas. For this reason, tables for maximum fan speeds at various exit-gas temperatures must be furnished to operating personnel.

Although the discharge air cirucit is not directly related to the actual production of cement, it nevertheless requires close attention from the operator. An operator must focus his attention upon the temperature of the gases at this location of the air circuit because either too high or low a temperature can be harmful to the dust-collecting equipment. The moisture and the sulfur content of the gases require that this temperature be controlled within a predetermined range. If the temperature falls below the dewpoint of the gases, the moisture can precipitate out in the baghouse or electrostatic precipitator, causing plug-up problems and chemical attack due to corrosion. Filter bags also have a given maximum-service temperature above which these bags can be damaged. Most baghouses require that the gases passing through do not exceed 285 C (545 F). The dewpoint of the gases varies depending on local prevailing conditions. Experience has shown that when the dust-collector inlet temperature is held above 160 C (320 F) there exists little danger of moisture precipitating out in the collector. In summary, it can be stated that most kiln systems have to operate the dust-collector inlet temperature within the range of 160–285 C (320–545 F) to prevent operating problems in the dust collector. It is important to remember that these temperatures apply to the *dust-collector inlet* and not to the kiln exit or preheater-tower exit-gas temperature. The reader might question the aforementioned dewpoint temperature for it is true that the dewpoint for kiln gases is normally between 50–90 C (120–195 F). Why then the adherence to the higher temperature given above? The answer is infiltration in the dust collector itself. Kiln gases are being further cooled inside these units by as much as 60 C (140 F) which can bring these gases close to the dewpoint even when they enter at a much higher temperature.

Another factor that has to be considered in respect to the discharge air circuit is the air inleakage between the kiln exit and the dust collector. This inleakage should be held to a minimum by effective sealing of all unnecessary openings. These openings diminish the induced draft fan (I.D. fan) capacity, require extra power to drive this fan, and could drop the gas temperature below the previously discussed, harmful dewpoint. It can be calculated that 25% infiltrated air at the discharge circuit can increase the horsepower requirements of the I.D. fan by an approximately equal amount of 25% which is a significant inefficiency in the system. Sealing doors and openings within the gas ducts is normally not difficult but, obtaining an effective kiln back-end seal is usually associated with high installation and maintenance costs. Nevertheless, these expenditures for effective seals can be financially justified by the resulting reduced horsepower requirements of the fan.

The amount of air infiltration at the discharge circuit can be measured by extracting and analyzing simultaneously the gases at the kiln exit and the dust-collector inlet. The percent infiltrated air is then calculated by Eq. (12-5):

$$I = \frac{OX_f - OX_o}{20.9 - OX_f} \times 100 \tag{12-5}$$

where:

I = percent of infiltrated air (by volume)
OX_f = percent O_2 at dust-collector inlet
OX_o = percent O_2 at kiln discharge.

Any other two sampling points can be selected in the system, such as dust-collector inlet/dust-collector outlet, to determine infiltration.

13.

Movement of Material Through the Kiln

13.1 KILN LOADING

The feed, as it advances through the kiln from the rear, through the several zones, and on to the cooler, does not move at a uniform speed, nor does it move in a straight line along the axis of the kiln. In the dehydration reach of the kiln behind the calcining zone, the feed first enters the chain section in which chains, attached loosely to the interior of the kiln, serve to agitate the feed to facilitate heat transfer. The speed with which the feed advances through this section is largely influenced by the chain pattern and density. The agitation and tumbling action cause considerable mixing of the feed constituents. Upon leaving the chain section the feed is raised part way up the arch as the kiln turns, following a path perpendicular to the kiln axis. At a certain point gravity causes the mass to slide down the kiln shell in a forward direction, and the movement is then repeated. In this manner the feed advances in a zig-zag path, each rise and slide advancing all parts of the feed a few inches. In contrast to the chain section, very little mixing takes place in the feed bed in this area.

The feed now enters the calcining zone, and the zig-zag course diminishes progressively as the feed bed travels down the kiln, because, with the evolution of carbon dioxide gas, the feed bed becomes partly fluidized and travels more rapidly, flowing in a manner somewhat akin to the flow of water, as well as continuing the rise-and-slide advancement. A slowdown occurs next in the burning zone. The feed, now transformed into a semi-liquid state, becomes sticky, starts to form clinkers, and undergoes a cascading action similar to that in the dehydration zone.

Naturally, there are irregularities that occur in this movement of feed through the kiln. Sometimes, when viewing the burning zone through the observation port, one can observe an onrush of dust waves that flush with considerable speed into the burning zone. These waves, moving much faster than the regularly advancing material bed, originate in the calcining zone because of erratic calcination resulting from some irregularity in kiln operation which caused the bed to become fluidized. They are much more frequent in small-diameter kilns with steep slopes than in large-diameter kilns with smaller inclination.

The opposite condition sometimes occurs in which material is retained for a prolonged period of time in a certain area of the kiln, caused by ring formation in the back of the burning zone or less frequently in the chain section. When any such ring breaks loose the kiln operator is confronted with a so-called "push" in which all the feed retained by the ring enters the burning zone along with fragments of the broken ring.

In both cases these reactions cause an uneven bed depth and constitute difficult operating conditions which can cause the kiln to go into an upset. Such upsets can occur in different magnitudes. Usually only small adjustments to kiln operation are necessary to maintain uniformity of operation of the kiln. At other times the condition may be so severe as to require more drastic measures, even to the extent of taking emergency procedures such as shutting down the kiln to protect the equipment. For these reasons it is useless to attempt to explain the steps necessary to counteract changes in feed-bed movement. Only experience and knowledge of each individual kiln will enable a kiln operator to know what kind of counteracting procedure should be carried out for any given condition. It should be emphasized, however, that it is extremely important to consider the after-effects an upset can have in the feed bed in the areas behind the burning zone. If the operator is content to take care only of the material in the burning zone during either a push or a light load, without anticipating what changes have taken place in the rear of the kiln, the kiln might be permitted to overreact to the opposite extreme. This could lead the kiln into a cycling condition that would require hours or even days to eliminate.

The mentioned zig-zag movement of the feed in the calcining zone is not conducive to good heat exchange between the gases and the feed. For lack of a tumbling action, the center of the feed bed experiences little contact with the gases whereas the particles on the surface get maximum exposure and thus maximum heat exchange. This action causes an uneven calcination within the cross section of the feed bed. Only shortly before the feed enters the burning zone will the increased tumbling action cause the "cold" core of the bed to be exposed to the gases. At that time, feed will then undergo its final and rapid calcination. This action is especially noticeable and pronounced in long wet- and dry-process kilns. Uneven calcination of the feed bed can often give an operator the false impression that the kiln-feed composition is too difficult to burn, i.e., the laboratory is mistakenly blamed for having prepared a tough-burning mix.

On wet- and dry-process kilns, lifters and cam linings are designed to promote mixing of the feed bed in the calcining zone and thus obtain a more uniform and better heat exchange for the bed as a whole. Many plants, however, experience higher and sometimes unacceptable dust generation as a result of these heat exchanger installations. A new type of lifter, called a disperser, has been tried on a few dry and wet kilns with good initial results. The intention of this disperser is to accomplish a stirring of the feed bed without causing the feed to be lifted into the gas stream. The dispersers are designed so that the feed slides across the uphill side and tumbles over a ledge on the downhill side. Operating data on these kilns has shown improved kiln stability, lower specific fuel consumption, and in one instance an increase in kiln output. Most significantly, the internal dust cycle and dust-waste rates do not increase noticeably after these dispersers are installed. On the negative side it must be mentioned that these dispersers are difficult to maintain and show a high wear and spall rate.

Under normal conditions, it is possible to compute, with reasonable accuracy, the time required for the feed to move through the kiln. Gibbs[1] proposes this equation:

$$T = \frac{11.4L}{NDS}$$

in which

T = travel time (min)
L = length of kiln (ft) or (m)
N = speed of kiln (revolutions per hour)
D = internal diameter of kiln (ft) or (m)
S = slope of kiln (ft/ft) or (m/m)

The constant 11.4 applies to cement kilns and is obtained from the relationship:

$$\text{constant} = 1.77 * \sqrt{\phi}$$

where:

ϕ is the average angle of repose for the material bed (assumed to be between 35 and 43 degrees for cement kilns). However, visual observation of the burning zone shows this angle to be more in the magnitude of 50 degrees. In the calcining zone, this angle is also considerably less due to the partial fluidization of the bed. Tests performed by the author on a rotating cylinder (1.0 rpm), having a smooth lining surface, showed an angle of repose of less than 20 degrees.

In Table 13.1 the retention time, feed-bed velocity, and percent kiln loading have been calculated to demonstrate, in theory, how these variables can change as the feed bed travels through the different zones down the kiln. The angles of repose and kiln-feed densities have been estimated to present, as closely as possible, actual kiln conditions.

Percent loading of the kiln is calculated as follows:

a) English system:

$$\% \text{ loading} = \frac{\dfrac{100(1.3888wf)}{d}}{\left(\dfrac{l}{t}\right)A}$$

b) Metric system:

$$\% \text{ loading} = \frac{\dfrac{100(0.69444wf)}{d}}{\left(\dfrac{l}{t}\right)A}$$

where:

w = kiln output (sh.tons/day) or (metric tons/day)
f = feed factor (lb feed/lb clinker)
d = feed density (lb/ft^3) or (kg/m^3)
l = kiln length (ft) or (m)
t = retention time (min)
A = effective cross-sectional area inside kiln (ft^2) or (m^2)

The percent loading of the kiln is a factor that is of considerable interest

TABLE 13.1

DRY-PROCESS KILN RETENTION TIME AND KILN LOADING

Kiln Data:

Kiln length: 525
Kiln I.D.: 14.50
Slope: 0.0365

Kiln speed: 72
Retention time (Gibbs formula): 157

	Preheat zone	10% calcined	20% calcined	40% calcined	80% calcined	Burning zone	Total and average values
Feed factor	1.58	1.522	1.464	1.348	1.116	1.00	1.383
Feed (lb)/clinker(t)	3160	3044	2928	2696	2232	2000	2765
Length of zone (ft)	200	80	60	50	40	95	525
Estimated angle of repose	45	27	24	21	18	50	36.42
Inside diameter (ft)	14.5	14.5	14.5	14	13.25	12.8	14.05
Estimated density (lb/ft^3)	75	73	70	65	60	92	75.10
Retention time (min)	65	20	14	12	9	37	157
Feed velocity (ft/min)	3.06	3.95	4.19	4.32	4.42	2.56	3.3
Loading (%)	10	7.7	7.3	7.5	7.3	7.9	8.6

to the engineer and operator. It is also a subject that can generate considerable disagreement among operators for one might prefer to run the kiln at higher speed and lower percent loading whereas another might prefer the opposite, namely, slow speed with a corresponding deeper bed. Operators can vary this loading by adjustment to:

a)   the kiln speed, or
b)   the feed rate.

Most kilns operate at 6–10% loading while kilns with a steep slope tend to operate at the higher end and kilns with shallower slopes on the lower end of this scale. The optimum setting can usually only be found by trial and error. To find this optimum setting, a kiln is usually operated for a few days at various percent loadings and its effect on kiln operating stability and efficiency investigated. The apparent "ideal" setting will then become the standard for all future operations on that kiln. More discussions on this subject follow later when kiln-speed and feed-rate controls are individually discussed.

13.2   KILN CAPACITY

A multitude of formulas have been advanced by various writers pertaining to the determination of the kiln capacity. Results obtained from these calculations can vary to a considerable extent. The choice of formulas narrows down to the question of what is meant by the word capacity. A kiln manufacturer, for example, will state the kiln capacity in terms of what is considered attainable, reasonable output rates, i.e., conservative. It is not unusual to exceed this stated capacity by as much as 10–15% for this author has consistently worked kilns that outproduced their rated capacity. Often, too, the operators are heard mentioning that a given kiln is capable of producing a lot more clinker but that either the I.D. fan, the cooler, or the raw-grinding capacity is a bottleneck. Then there are numerous cases where operators might be of the opinion, sometimes rightfully so, that the kiln is being "force" fired, i.e., being crammed with feed to such an extent that it causes undue operating problems.

G. Martin[1] made a strong point in 1937 when he determined that the kiln capacity is not a function of the overall kiln volume but that the kiln diameter is the governing factor in determining how much a kiln is capable of producing. At the time when he wrote his book, there were no pre-calciner kilns in existence to give him the proof that his reasoning was correct. The formulas he advanced for determining the kiln capacity have also become outdated with the installation of induced draft fans on kilns because his calculations were based on natural kiln drafts.

From an operator's viewpoint, kiln capacity is that limit above which gas velocities become so high that kiln feed is easily lifted into the gas stream, i.e., the internal dust cycle becomes excessively large and starts to interfere with the stability of the operation. These conditions are reached when the kiln-gas velocity exceeds a speed of approximately 9 m/s (30 ft/s). A kiln that operates in the calcining zone with a higher velocity than indicated above is considered to be "force fired." Since this is the so-called threshold limit, kilns usually are operated at output rates that are 5–10% below the velocity-calculated capacity.

The kiln-gas velocity is calculated by the well-known formula:

$$\text{Velocity} = \text{volume/area}$$

and is thus influenced by the following:

a) the effective cross-sectional area of the kiln
   - kiln diameter inside lining
   - percent loading of kiln
   - inserts such as chains, lifters
   - coating and ring formation
b) the volume of gas passing through this cross section
   - fuel firing rate (exclude fuel to precalciner)
   - gases emanating from the feed
   - temperature of the gas
   - amount of excess air in kiln

With the help of a computer the capacity tables (Tables 13.2–13.8) have been developed to show the threshold (limit) output rates for various kiln systems. The tables are printed in both the English and Metric systems of units, hence, care has to be taken to read the appropriate tables. To simplify computations, the following constants have been used in the development of these tables:

---

[1] G. Martin, *Chemical engineering and thermodynamics applied to rotary cement kilns.* London: Technical Press Ltd.

	English	Metric
	English	*Metric*
Fuel heating value	12,600 Btu/lb	29,288 kJ/kg
Combustion gas	11.6 lb/lb coal	11.6 kg/kg coal
(@ 5% excess air)		
Gas from feed	0.534 lb/lb clinker	0.534 kg/kg clinker
(@ 100% calcination)		
Gas density (std. cond.)	0.09 lb/ft^3	1.44 kg/Nm^3
Loading in calcining zone (%)	9%	9%
Gas temperature (dry and wet kiln)	1600 F	871 C
(preheater kiln)	1900 F	1038 C
(precalciner kiln)	2150 F	1177 C

These tables can be used to answer the questions if a given kiln is being "force fired" or if the output is considerably below its capacity. If a kiln output is higher than indicated in the table, and if this kiln is plagued with instability, an effort should be made to lower the production rate. On the other hand, when a given kiln consistently produces less than 80% of the indicated capacity, efforts should be undertaken to raise the output provided the auxiliary equipment can handle this increase.

TABLE 13.2
WET-PROCESS KILN CAPACITY

English units (sh.t/h)

9% loading in calcining zone

Kiln I.D.	MBtu/t clinker							
	4.8	5.0	5.2	5.4	5.6	5.8	6.0	6.2
10	30.2	29.3	28.3	27.5	26.7	25.9	25.2	24.5
11	36.6	35.4	34.3	33.2	32.3	31.3	30.5	29.6
12	43.5	42.1	40.8	39.6	38.4	37.3	36.2	35.3
13	51.1	49.4	47.9	46.4	45.1	43.8	42.5	41.4
14	59.3	57.3	55.5	53.9	52.3	50.8	49.3	48.0
15	68.0	65.8	63.8	61.8	60.0	58.3	56.6	55.1
16	77.4	74.9	72.5	70.3	68.3	66.3	64.4	62.7
17	87.4	84.6	81.9	79.4	77.1	74.8	72.7	70.8
18	98.0	94.8	91.8	89.0	86.4	83.9	81.6	79.3
19	109.2	105.6	102.3	99.2	96.2	93.5	90.9	88.4
20	121.0	117.0	113.4	109.9	106.6	103.6	100.7	98.0
21	133.4	129.0	125.0	121.2	117.6	114.2	111.0	108.0

Metric units (metric t/h clinker)

9% loading in calcining zone

Kiln I.D.	kJ/kg clinker							
	5579	5811	6044	6276	6508	6741	6973	7206
3.049	27.4	26.5	25.7	24.9	24.2	23.5	22.8	22.2
3.354	33.2	32.1	31.1	30.2	29.3	28.4	27.6	26.9
3.659	39.5	38.2	37.0	35.9	34.8	33.8	32.9	32.0
3.963	46.4	44.9	43.4	42.1	40.9	39.7	38.6	37.5
4.268	53.8	52.0	50.4	48.9	47.4	46.0	44.8	43.5
4.573	61.7	59.7	57.8	56.1	54.4	52.9	51.4	50.0
4.878	70.2	68.0	65.8	63.8	61.9	60.1	58.5	56.9
5.183	79.3	76.7	74.3	72.0	69.9	67.9	66.0	64.2
5.488	88.9	86.0	83.3	80.8	78.4	76.1	74.0	72.0
5.793	99.0	95.8	92.8	90.0	87.3	84.8	82.4	80.2
6.098	109.7	106.2	102.8	99.7	96.8	94.0	91.3	88.9
6.402	121.0	117.1	113.4	109.9	106.7	103.6	100.7	98.0

TABLE 13.3

DRY-PROCESS KILN CAPACITY

English units (sh.t/h)

9% loading in calcining zone

Kiln I.D.	3.8	3.9	4	4.1	4.2	4.3	4.4	4.5
				MBtu/t clinker				
10	36.3	35.6	34.9	34.3	33.6	33.0	32.4	31.8
11	44.0	43.1	42.3	41.5	40.7	39.9	39.2	38.5
12	52.3	51.3	50.3	49.3	48.4	47.5	46.7	45.9
13	61.4	60.2	59.0	57.9	56.8	55.8	54.8	53.8
14	71.2	69.8	68.5	67.2	65.9	64.7	63.5	62.4
15	81.8	80.1	78.6	77.1	75.7	74.3	72.9	71.6
16	93.0	91.2	89.4	87.7	86.1	84.5	83.0	81.5
17	105.0	102.9	100.9	99.0	97.2	95.4	93.7	92.0
18	117.7	115.4	113.2	111.0	108.9	107.0	105.0	103.2
19	131.2	128.6	126.1	123.7	121.4	119.2	117.0	115.0
20	145.3	142.5	139.7	137.1	134.5	132.0	129.7	127.4
21	160.2	157.1	154.0	151.1	148.3	145.6	143.0	140.4

Metric units (metric t/h)

9% loading in calcining zone

Kiln I.D.	kJ/kg clinker							
	4416	4533	4649	4765	4881	4998	5114	5230
3.049	33.0	32.3	31.7	31.1	30.5	29.9	29.4	28.9
3.354	39.9	39.1	38.3	37.6	36.9	36.2	35.6	35.0
3.659	47.5	46.5	45.6	44.8	43.9	43.1	42.3	41.6
3.963	55.7	54.6	53.6	52.5	51.6	50.6	49.7	48.8
4.268	64.6	63.3	62.1	60.9	59.8	58.7	57.6	56.6
4.573	74.2	72.7	71.3	69.9	68.6	67.4	66.2	65.0
4.878	84.4	82.7	81.1	79.6	78.1	76.7	75.3	74.0
5.183	95.3	93.4	91.6	89.8	88.2	86.5	85.0	83.5
5.488	106.8	104.7	102.7	100.7	98.8	97.0	95.3	93.6
5.793	119.0	116.7	114.4	112.2	110.1	108.1	106.2	104.3
6.098	131.9	129.3	126.8	124.3	122.0	119.8	117.6	115.6
6.402	145.4	142.5	139.7	137.1	134.5	132.1	129.7	127.4

TABLE 13.4

SUSPENSION PREHEATER KILN CAPACITY (20% calcined in cyclones)

English units (sh.t/h)

80% calcination done in kiln
9% loading in calcining zone

Kiln I.D.	MBtu/t clinker							
	2.8	2.9	3	3.1	3.2	3.3	3.4	3.5
10	42.2	41.1	40.0	39.1	38.1	37.2	36.3	35.5
11	51.1	49.7	48.5	47.3	46.1	45.0	44.0	43.0
12	60.8	59.2	57.7	56.2	54.9	53.6	52.3	51.2
13	71.3	69.5	67.7	66.0	64.4	62.9	61.4	60.0
14	82.7	80.5	78.5	76.5	74.7	72.9	71.2	69.6
15	94.9	92.5	90.1	87.9	85.7	83.7	81.8	79.9
16	108.0	105.2	102.5	100.0	97.6	95.3	93.1	91.0
17	122.0	118.8	115.7	112.9	110.1	107.5	105.0	102.7
18	136.7	133.2	129.8	126.5	123.5	120.6	117.8	115.1
19	152.3	148.4	144.6	141.0	137.6	134.3	131.2	128.3
20	168.8	164.4	160.2	156.2	152.4	148.8	145.4	142.1
21	186.1	181.2	176.6	172.2	168.1	164.1	160.3	156.7

Metric units (metric t/h)

80% calcination done in kiln
9% loading in calcining zone

Kiln I.D.	kJ/kg clinker							
	3254	3370	3487	3603	3719	3835	3952	4068
3.049	38.3	37.3	36.3	35.4	34.6	33.8	33.0	32.2
3.354	46.3	45.1	44.0	42.9	41.8	40.8	39.9	39.0
3.659	55.1	53.7	52.3	51.0	49.8	48.6	47.5	46.4
3.963	64.7	63.0	61.4	59.9	58.4	57.0	55.7	54.5
4.268	75.0	73.1	71.2	69.4	67.8	66.2	64.6	63.2
4.573	86.1	83.9	81.7	79.7	77.8	75.9	74.2	72.5
4.878	98.0	95.4	93.0	90.7	88.5	86.4	84.4	82.5
5.183	110.6	107.7	105.0	102.4	99.9	97.6	95.3	93.1
5.488	124.0	120.8	117.7	114.8	112.0	109.4	106.8	104.4
5.793	138.2	134.6	131.2	127.9	124.8	121.9	119.0	116.4
6.098	153.1	149.1	145.3	141.7	138.3	135.0	131.9	128.9
6.402	168.8	164.4	160.2	156.3	152.5	148.9	145.4	142.1

TABLE 13.5

SUSPENSION PREHEATER KILN CAPACITY (40% calcined in cyclones)

English units (sh.t./h)

60% calcination done in kiln
9% loading in calcining zone

Kiln I.D.	MBtu/t clinker							
	2.8	2.9	3.0	3.1	3.2	3.3	3.4	3.5
10	45.0	43.7	42.6	41.4	40.4	39.4	38.4	37.5
11	54.4	52.9	51.5	50.1	48.9	47.6	46.5	45.4
12	64.8	63.0	61.3	59.7	58.1	56.7	55.3	54.0
13	76.0	73.9	71.9	70.0	68.2	66.5	64.9	63.4
14	88.2	85.7	83.4	81.2	79.1	77.2	75.3	73.5
15	101.2	98.4	95.8	93.2	90.9	88.6	86.4	84.4
16	115.2	112.0	109.0	106.1	103.4	100.8	98.3	96.0
17	130.0	126.4	123.0	119.8	116.7	113.8	111.0	108.4
18	145.8	141.7	137.9	134.3	130.8	127.6	124.4	121.5
19	162.4	157.9	153.7	149.6	145.8	142.1	138.7	135.3
20	180.0	175.0	170.3	165.8	161.5	157.5	153.6	150.0
21	198.4	192.9	187.7	182.8	178.1	173.6	169.4	165.3

Metric units (metric t/h)

60% calcination done in kiln
9% loading in calcining zone

Kiln I.D.	kJ/kg clinker							
	3254	3370	3487	3603	3719	3835	3952	4068
3.049	40.8	39.7	38.6	37.6	36.6	35.7	34.8	34.0
3.354	49.4	48.0	46.7	45.5	44.3	43.2	42.2	41.2
3.659	58.8	57.2	55.6	54.1	52.7	51.4	50.2	49.0
3.963	69.0	67.1	65.3	63.5	61.9	60.4	58.9	57.5
4.268	80.0	77.8	75.7	73.7	71.8	70.0	68.3	66.7
4.573	91.9	89.3	86.9	84.6	82.4	80.4	78.4	76.5
4.878	104.5	101.6	98.9	96.2	93.8	91.4	89.2	87.1
5.183	118.0	114.7	111.6	108.7	105.9	103.2	100.7	98.3
5.488	132.3	128.6	125.1	121.8	118.7	115.7	112.9	110.2
5.793	147.4	143.3	139.4	135.7	132.2	128.9	125.8	122.8
6.098	163.3	158.8	154.5	150.4	146.5	142.9	139.4	136.1
6.402	180.0	175.0	170.3	165.8	161.5	157.5	153.7	150.0

TABLE 13.6
PRECALCINER KILN CAPACITY (50% fuel to precalciner)

English units (sh.t/h)

0.1% calcination done in kiln
0.5% of total heat input into kiln
9% loading in calcining zone

Kiln I.D.	MBtu/t input into kiln							
	1.35	1.40	1.45	1.50	1.55	1.60	1.65	1.70
10	107.3	103.8	100.5	97.4	94.4	91.7	89.1	86.6
11	129.8	125.6	121.6	117.8	114.3	110.9	107.8	104.8
12	154.5	149.4	144.7	140.2	136.0	132.0	128.3	124.7
13	181.4	175.4	169.8	164.5	159.6	154.9	150.5	146.4
14	210.3	203.4	196.9	190.8	185.1	179.7	174.6	169.8
15	241.5	233.5	226.0	219.0	212.5	206.3	200.4	194.9
16	274.7	265.7	257.2	249.2	241.7	234.7	228.0	221.8
17	310.1	299.9	290.3	281.3	272.9	264.9	257.4	250.4
18	347.7	336.2	325.5	315.4	305.9	297.0	288.6	280.7
19	387.4	374.6	362.7	351.4	340.9	331.0	321.6	312.7
20	429.2	415.1	401.8	389.4	377.7	366.7	356.3	346.5
21	473.2	457.6	443.0	429.3	416.4	404.3	392.9	382.0

Metric units (metric t/h)

0.1% calcination done in kiln
0.5% of total heat input into kiln
9% loading in calcining zone

Kiln I.D.	kJ/kg input into kiln							
	1569	1627	1685	1743	1801	1860	1918	1976
3.049	97.4	94.1	91.1	88.3	85.7	83.2	80.8	78.6
3.354	117.8	113.9	110.3	106.9	103.7	100.6	97.8	95.1
3.659	140.2	135.6	131.2	127.2	123.4	119.8	116.4	113.2
3.963	164.5	159.1	154.0	149.3	144.8	140.6	136.6	132.8
4.268	190.8	184.5	178.6	173.1	167.9	163.0	158.4	154.0
4.573	219.0	211.8	205.1	198.7	192.7	187.1	181.8	176.8
4.878	249.2	241.0	233.3	226.1	219.3	212.9	206.9	201.2
5.183	281.3	272.1	263.4	255.2	247.6	240.4	233.6	227.1
5.488	315.4	305.0	295.3	286.1	277.6	269.5	261.8	254.6
5.793	351.4	339.9	329.0	318.8	309.3	300.2	291.7	283.7
6.098	389.4	376.6	364.5	353.3	342.7	332.7	323.3	314.4
6.402	429.3	415.2	401.9	389.5	377.8	366.8	356.4	346.6

TABLE 13.7

PRECALCINER KILN CAPACITY (45% fuel to precalciner)

English units (sh.t/h)

0.2% calcination done in kiln
0.55% of total heat input into kiln
9% loading in calcining zone

Kiln I.D.	MBtu/t input into kiln							
	1.485	1.540	1.595	1.650	1.705	1.760	1.815	1.870
10	91.6	88.8	86.1	83.6	81.2	79.0	76.9	74.8
11	110.9	107.4	104.2	101.1	98.3	95.6	93.0	90.6
12	131.9	127.8	124.0	120.4	117.0	113.7	110.7	107.8
13	154.8	150.0	145.5	141.3	137.3	133.5	129.9	126.5
14	179.6	174.0	168.8	163.8	159.2	154.8	150.6	146.7
15	206.2	199.8	193.7	188.1	182.7	177.7	172.9	168.4
16	234.6	227.3	220.4	214.0	207.9	202.2	196.7	191.6
17	264.8	256.6	248.8	241.6	234.7	228.2	222.1	216.3
18	296.9	287.6	279.0	270.8	263.1	255.9	249.0	242.5
19	330.8	320.5	310.8	301.8	293.2	285.1	277.4	270.2
20	366.5	355.1	344.4	334.4	324.9	315.9	307.4	299.4
21	404.1	391.5	379.7	368.6	358.2	348.3	338.9	330.1

Metric units (metric t/h)

0.2% calcination done in kiln
0.55% of total heat input into kiln
9% loading in calcining zone

Kiln I.D.	kJ/kg input into kiln							
	1726	1790	1854	1918	1982	2046	2109	2173
3.049	83.1	80.5	78.1	75.8	73.7	71.6	69.7	67.9
3.354	100.6	97.5	94.5	91.8	89.2	86.7	84.4	82.2
3.659	119.7	116.0	112.5	109.2	106.1	103.2	100.4	97.8
3.963	140.5	136.1	132.0	128.2	124.5	121.1	117.8	114.7
4.268	162.9	157.9	153.1	148.6	144.4	140.4	136.7	133.1
4.573	187.0	181.2	175.8	170.6	165.8	161.2	156.9	152.8
4.878	212.8	206.2	200.0	194.1	188.6	183.4	178.5	173.8
5.183	240.2	232.8	225.8	219.2	212.9	207.1	201.5	196.2
5.488	269.3	261.0	253.1	245.7	238.7	232.1	225.9	220.0
5.793	300.1	290.8	282.0	273.8	266.0	258.6	251.7	245.1
6.098	332.5	322.2	312.5	303.3	294.7	286.6	278.9	271.6
6.402	366.6	355.2	344.5	334.4	324.9	316.0	307.5	299.4

TABLE 13.8

PRECALCINER KILN CAPACITY (40% fuel to precalciner)

English units (sh.t/h)

0.3% calcination done in kiln
0.6% of total heat input into kiln
9% loading in calcining zone

Kiln I.D.	MBtu/t input into kiln							
	1.62	1.68	1.74	1.80	1.86	1.92	1.98	2.04
10	79.9	77.6	75.3	73.2	71.2	69.4	67.6	65.9
11	96.7	93.9	91.2	88.6	86.2	83.9	81.8	79.7
12	115.1	111.7	108.5	105.5	102.6	99.9	97.3	94.9
13	135.1	131.1	127.3	123.8	120.4	117.2	114.2	111.3
14	156.7	152.0	147.7	143.5	139.6	136.0	132.4	129.1
15	179.9	174.5	169.5	164.8	160.3	156.1	152.0	148.2
16	204.6	198.6	192.9	187.5	182.4	177.6	173.0	168.6
17	231.0	224.2	217.7	211.7	205.9	200.5	195.3	190.4
18	259.0	251.3	244.1	237.3	230.8	224.7	218.9	213.4
19	288.6	280.0	272.0	264.4	257.2	250.4	243.9	237.8
20	319.7	310.3	301.4	293.0	285.0	277.5	270.3	263.5
21	352.5	342.1	332.3	323.0	314.2	305.9	298.0	290.5

Metric units (metric t/h)

0.3% calcination done in kiln
0.6% of total heat input into kiln
9% loading in calcining zone

Kiln I.D.	kJ/kg input into kiln							
	1883	1953	2022	2092	2162	2231	2301	2371
3.049	72.5	70.4	68.4	66.4	64.6	62.9	61.3	59.8
3.354	87.7	85.2	82.7	80.4	78.2	76.1	74.2	72.3
3.659	104.4	101.3	98.4	95.7	93.1	90.6	88.3	86.1
3.963	122.6	118.9	115.5	112.3	109.2	106.3	103.6	101.0
4.268	142.1	137.9	134.0	130.2	126.7	123.3	120.2	117.1
4.573	163.2	158.3	153.8	149.5	145.4	141.6	137.9	134.5
4.878	185.6	180.2	175.0	170.1	165.5	161.1	156.9	153.0
5.183	209.6	203.4	197.5	192.0	186.8	181.9	177.2	172.7
5.488	235.0	228.0	221.5	215.3	209.4	203.9	198.6	193.6
5.793	261.8	254.0	246.7	239.9	233.3	227.2	221.3	215.8
6.098	290.1	281.5	273.4	265.8	258.5	251.7	245.2	239.1
6.402	319.8	310.3	301.4	293.0	285.0	277.5	270.4	263.6

TABLE 13.9

PRECALCINER KILN CAPACITY (30% fuel to precalciner)

English units (sh.t/h)

0.4% calcination done in kiln
0.7% of total heat input into kiln
9% loading in calcining zone

Kiln I.D.	MBtu/t input into kiln							
	1.89	1.96	2.03	2.10	2.17	2.24	2.31	2.38
10	66.8	64.9	63.1	61.4	59.7	58.2	56.7	55.3
11	80.9	78.5	76.3	74.2	72.3	70.4	68.6	66.9
12	96.2	93.5	90.8	88.4	86.0	83.8	81.7	79.7
13	112.9	109.7	106.6	103.7	100.9	98.3	95.8	93.5
14	131.0	127.2	123.6	120.3	117.1	114.0	111.2	108.4
15	150.4	146.0	141.9	138.1	134.4	130.9	127.6	124.5
16	171.1	166.1	161.5	157.1	152.9	148.9	145.2	141.6
17	193.1	187.6	182.3	177.3	172.6	168.1	163.9	159.9
18	216.5	210.3	204.4	198.8	193.5	188.5	183.7	179.2
19	241.3	234.3	227.7	221.5	215.6	210.0	204.7	199.7
20	267.3	259.6	252.3	245.4	238.9	232.7	226.8	221.3
21	294.7	286.2	278.2	270.6	263.4	256.6	250.1	243.9

Metric units (metric t/h)

0.4% calcination done in kiln
0.7% of total heat input into kiln
9% loading in calcining zone

Kiln I.D.	kJ/kg input into kiln							
	2197	2278	2359	2441	2522	2603	2685	2766
3.049	60.6	58.9	57.2	55.7	54.2	52.8	51.4	50.2
3.354	73.4	71.2	69.2	67.4	65.6	63.9	62.3	60.7
3.659	87.3	84.8	82.4	80.2	78.0	76.0	74.1	72.3
3.963	102.5	99.5	96.7	94.1	91.6	89.2	86.9	84.8
4.268	118.8	115.4	112.2	109.1	106.2	103.4	100.8	98.4
4.573	136.4	132.5	128.8	125.2	121.9	118.8	115.8	112.9
4.878	155.2	150.7	146.5	142.5	138.7	135.1	131.7	128.5
5.183	175.2	170.2	165.4	160.9	156.6	152.5	148.7	145.0
5.488	196.4	190.8	185.4	180.3	175.6	171.0	166.7	162.6
5.793	218.9	212.5	206.6	200.9	195.6	190.5	185.7	181.2
6.098	242.5	235.5	228.9	222.6	216.7	211.1	205.8	200.7
6.402	267.4	259.6	252.4	245.5	238.9	232.8	226.9	221.3

Part II

Kiln Operating Procedures

14.

Kiln Operating and Control Methods

Methods of kiln control vary from plant to plant, and even between different operators, because each operator is apt to have his own ideas as to how to proceed when confronted with any given situation. Kiln control, however, must be a continuous around-the-clock matter, hence it is necessary for all operators on all shifts to operate the kiln in the same manner. This in turn means that all operators should be trained in the same principles of kiln operation.

There is nothing more destructive to operating stability than the changing of controller setpoints en masse during shift changes. Unless an emergency exists or an obvious change must be made, it is not possible for an operator to assess the need for controller-setting changes in the first few minutes after coming on duty. A difference in settings from the previous day is not necessarily an indication to change these settings back to where they were 24 h before. An operator should allow 20 min at the onset of the shift to first observe the process before deciding that a change is indeed warranted.

14.1 CLINKER-BURNING TECHNIQUES

There are three common techniques for burning clinker in a rotary kiln:

a) Maintain a constant kiln speed, and vary the fuel rate to counteract temperature changes in the burning zone.

   b) Maintain a constant fuel rate, and vary the kiln speed to hold the burning-zone temperature at the desired level.
   c) Vary the kiln speed, the fuel rate, or both, to maintain the desired burning-zone temperature.

These techniques have one error in common; they show concern only for the burning-zone temperature. Unfortunately, many kiln operators think that this is good enough, reasoning that, as long as good clinker is produced, what more is necessary? The fallacy of this reasoning lies in the fact that ideal stable kiln conditions can be obtained faster and more economically when equal consideration is given to all zones in the kiln and not the burning zone alone. Drying and calcining of the feed must be considered before one can consider making clinker. The process of clinker burning, and therefore the process of rotary-kiln control, starts not at the place where the feed enters the burning zone, but at the point where the feed enters the kiln. In preheater and precalciner kilns this applies to the point where the feed is given to the top stage of the preheater cyclones.

The technique described in this book can be summarized as follows:

VARY THE KILN SPEED, THE FUEL RATE, AND THE INDUCED DRAFT-FAN SPEED IN ANY COMBINATION TO MAINTAIN THE PROPER BURNING-ZONE TEMPERATURE AND A CONSTANT KILN BACK-END TEMPERATURE FOR A GIVEN FEED RATE.

This is sometimes referred to as "burning a kiln from the rear," the most reliable technique that fulfills the four fundamental rules of clinker burning which, in order of priority, are:

   a)   Protection of the equipment and personnel at all times.
   b)   Production of well-burned clinker.
   c)   Continuous stable kiln operation.
   d)   Maximum production with maximum fuel efficiency.

   a) Protection of Equipment and Personnel. Safe operation must have first priority at all times; the operator should never allow himself to bypass this rule. Red grates in the cooler, red spots on the kiln shell, or overheated chains can do much more damage than the production of a few barrels of bad clinker, if bringing the equipment trouble under control would happen to result in the production of a small amount of poor clinker. It is usually during times of unusual and severe upset of the operating conditions that an operator will most likely forget this fundamental rule. This is because of the attention understandably being given in such instances to leading the kiln back to normal operating conditions. But, it is precisely during these "tight" moments that the shift in attention toward the safety of the equipment and fellow workers must be made. The majority of accidents happen when the kiln is in an unusual operating condition such as during kiln starts, stops, and upsets. Safety around a rotary kiln being an important and integral part of kiln operation is covered in greater details in Chapter 26.

b) Production of a Well-Burned Clinker. A well-burned clinker is a clinker that is neither underburned nor overburned, has been properly cooled, and possesses the correct free-lime content and the desired liter weight.

c) Continuous Stable Kiln Operation. Continuous operation should always have priority over maximum production. More can be gained by having emphasis on continuous operation rather than by pushing the kiln to peak production at the expense of periodical kiln upsets. Stable kiln operation is the key to long refractory life, high fuel efficiency, and uniform quality clinker.

The term "stable kiln condition" means the condition in which only very small changes, or no changes at all, have to be made to the control variables to hold the kiln in a state of equilibrium. It is the kiln operator's duty to obtain stable conditions and not be satisfied until this goal has been achieved. The capability of a kiln operator is measured not so much by the length of time a stable kiln condition is maintained but by how well kiln upsets are handled, and how much skill is used in leading the kiln back to stable conditions after an upset.

Stable operation is also identified by the recording charts, most noteworthy being the kiln speed, and burning-zone and back-end temperatures (all deviating very little over a long period of time). Upset conditions are marked by large changes in kiln speed and, when these speed changes occur in frequent intervals, e.g., every 2 h, the kiln is in a cycling condition. Working a kiln out of a cycle, i.e., breaking a cycle, is the most challenging task a kiln operator faces (see Chapter 25). This is the time when kiln burning becomes more of an art than a routine. Such kiln upsets can happen not only on dry- and wet-process kilns but, although less severe and less frequent, also on preheater and precalciner kilns. Even kilns with sophisticated computer and automatic control are not immune to these cycling conditions.

d) *Maximum Production with Maximum Fuel Efficiency.* Only after the three previous requirements have been met can one try to raise production and start to concentrate on the details of fuel efficiency. Any move made in this direction should be done very slowly in order to avoid upsetting the prevailing stable kiln condition. Large kilns usually operate better in the upper range of their rated production rate, hence production increases can help to stabilize the operation if the kiln has been operating somewhat below its rated capacity.

14.2 MANUAL CONTROL

Controlling a kiln in the conventional manner, as was done on most kilns twenty years ago, consisted of observing the operation and making a manual adjustment on the control panel based upon that observation. One of the many possible rotary kiln control functions is looking into the burning zone and noticing a change in temperature which requires an adjustment in the fuel rate. Experience tells the kiln operator that he can neither turn the fuel valve wide open nor fully closed, but must base the required fuel change on how much the temperature has changed. A small variation in temperature requires only a small adjustment, while a large temperature change necessitates a correspondingly large modification in the amount of fuel admitted into the kiln.

The next factor to take into consideration is the limit to which the burning-zone temperature can drop or increase without requiring more drastic action than adjustment to the fuel rate itself. Too much fuel addition, although justified by the existing burning-zone temperature, could lead later on to dangerous overheating or incomplete combustion. Likewise, cutting the fuel rate back futher than an established limit might lead the kiln into a severe upset condition. Before any adjustment is made, the operator must consider the limit to which the burning-zone temperature and the fuel rate can be permitted to go without leading the kiln into an upset.

After an adjustment has been made to the fuel rate, the next question is how long the fuel rate can be left at the new setting until it has to be adjusted again. A certain time will transpire from the moment the fuel rate is adjusted until the burning zone reacts to this change and returns to its target temperature again. However, when the temperature has reacted and starts to turn in the other direction, at a given point another fuel-rate adjustment must be made so that the temperature does not overshoot its target. In other words, the temperature should not move from one extreme to another but should return and level off at the target. This is the essence of manual process control and the thought process required of the operator.

By now it should be obvious that the skill and experience of the operator play an all-important part in this type of control. His judgment and decisions are the main factors governing the degree of stability the kiln operation will obtain. If one considers the fact that an operator often has to exercise control over not one but sometimes two and more kilns at the same time and that each kiln has a multitude of controls such as described above, it is understandable that many of these fine aspects of control can be overlooked. Every kiln operator, regardless of how long he has been on the job, knows that an error in judgment can sometimes be made or a wrong decision can be reached. Often a decision could not be made strictly according "to the book" at the time it had to be made because of some uncertainty in the process not apparent to the operator. The only thing known at the time could have been that some adjustment was necessary. This adjustment may be shown on the strip-chart recording a few hours later to be completely contrary to what should have been done. Nobody feels worse than the operator himself when such evidence shows up on the strip chart. Thus, it is easy to say afterwards that something different should have been done, but it is far more difficult to make the right decision at the time when some uncertainty existed and in an immediate decision had to be made.

Manual control undoubtedly demands a heavy work load from the operator because most of these manual adjustments are repetitious in nature. Fortunately, the cement industry has experienced a phenomenal revolution in automatic process control in the past 20 years. This has made the manually controlled kilns more of an exception and it is expected that, a few years hence, they will become a rarity.

A process as complicated as the operation of a rotary kiln necessitates a multitude of instruments and controllers. It would be impossible to control a kiln without the help of instruments. Not too long ago, kiln control was limited to measuring temperatures, pressures, and flow rates, transmitting the information to a recorder on the firing floor, then leaving actual control of the variables in the hands of the kiln operator. It was not unusual to observe an operator going from burner hood porthole to control panel and back again, making almost continuous adjustments to the controls. Understandably, operators became uneasy when the reactions didn't come out exactly as expected, as individual human error played a large part in this method of operating.

14.3 AUTOMATIC CONTROL

Today's kilns are not only equipped with more and better controllers, but the computer is also slowly taking over the controls of entire kiln systems. Automation has become the standard in the cement industry, performing the work the kiln operator had to do himself in the past. Concurrent with this revolution, there also appeared new terminologies, better known as computer jargon that the kiln operator must become familiar with. During the early stage, when automation was still in its infancy, there were numerous workers that showed apprehension and sometimes outright objection to these developments. Advances made in control technology over the past few years have mostly eliminated this initial apprehension. Computer systems have been "humanized," made more simple to understand, and control capabilities have become so reliable and advanced that even the most experienced kiln operators accept them as a blessing. It has definite advantages over pacing the burner floor for eight hours especially during cold winter and hot summer days. Control-room operators also have an ally in the computer engineer who can be of assistance when operating problems have to be resolved.

It would defeat the purpose of this book to discuss in details the numerous types of computer systems that are being used today in the cement industry. Little benefit could be derived by discussing the control concepts of any particular kiln, because each kiln has, to some extent, its own design and its own automatic control idiosyncracy.

15.

Instrumentation

15.1 AUTOMATIC CONTROLLERS

Having automatic controllers and computers to handle part or all of the kiln operation requires that the operator acquires a basic knowledge of the workings of an automatic controller. There is a specific need to familiarize an operator with this subject, for nobody else has a better opportunity to observe the results of an automatic controller in actual operation than he does. Automatic process control has a better chance of succeeding and will yield better results if the experience, observations, and suggestions of the operators are considered when the automatic control is initiated or adjusted.

In this chapter, some of the apparently difficult-to-understand principles of process control are explained in simple terms, using as an example the control of the temperature in the burning zone. The fact that repeated reference is made to the burning-zone temperature and the fuel rate does not mean that the described fundamentals are valuable for this example alone. These fundamentals are applicable almost universally to any kind of process control and the proper relationship between the example given here and a specific process under investigation should be easily recognized.

Regardless of whether a kiln is manual, automatic, or fully computer-controlled, trends of the more important process variables are either recorded on a recording chart or displayed on the CRT. Fig. 15.2 is a portion of a strip recorder chart similar to the one being examined in Fig. 15.3. This chart also shows the effect of "process noise," the normal momentary fluctuations (but not changes) in the variable being recorded. Line A is virtually free of process noise, but Line B suffers from conspicuous noise.

A recorder, by itself, only records what the input variable tells it to record. In other words, a recorder merely takes a visible record of a process

Fig. 15.1 Instrumentation for a modern cement mill consists of scores of recorders, controllers, and indicators, as shown in this view of a portion of a central control room. (*The Foxboro Co.*)

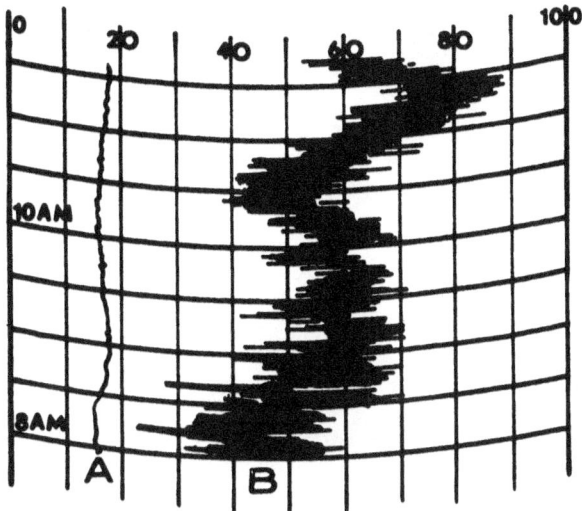

Fig. 15.2 Two variables are traced on this strip chart. Line A represents the grate speed in the cooler. Speed is quite uniform and the curve is free of process noise. Line B, percentage of oxygen in the kiln exit gas, shows a large increase in oxygen starting about 10:15 am, reaching a maximum at 11:25, then decreasing. Maximum span for oxygen content is from 0–5%; span reaches from 15–70% of chart (0.75-3.5% oxygen). Oxygen curve shows considerable process noise.

variable such as pressure, flow, temperature, or speed.

Under automatic control a *controller* in response to input signals, makes changes in the output variable, or control means, a process function that maintains the input variable within certin established limits. These changes or adjustments to the output variable can also be recorded (Fig. 15.4). For example, in response to variations in burning-zone temperature, the controller makes changes in the rate of fuel consumption.

An integral part of the controller system is the *transmitter*. Each of the two flow transmitters shown in Fig. 15.5 sends electrical signals to the recording controller which records and controls the flow in accordance with instructions that were programmed into the controller. A schematic diagram of instrumentation for a kiln and cooler is given in Fig. 15.6.

Automatic controllers are installed for the specific purpose of improving overall control of a process and eliminating costly human errors

Fig. 15.3   A strip chart from the process recorder in the instrument panel is being examined by the operator.   Two controllers are located immediately to the right of the recorder. (*The Foxboro Co.*)

that can occur when a process is manually controlled. When a controller meets these objectives, the work of the operator is made considerably easier (see Fig. 15.1).

The easy way out for an operator, when things do not go according to plan, is to take control manually and say that the controller is no good. The better approach, and the only one which can lead to satisfactory end result, is to find out *why* the controller does not do the job as it originally was designed to do. A serious effort in this direction will in most cases yield an answer.

When manual control is to be replaced by an automatic control, reliable instruments (in the following example, a pyrometer) have to be found and properly installed in order that a true measurement of the process variable can be obtained. Obtaining this correct value or process trend is an absolute requirement in automatic control. If this is not done, the effort to achieve automatic control is doomed to fail. Simply stated, the question becomes: How is it possible to control any variable, for example, a level

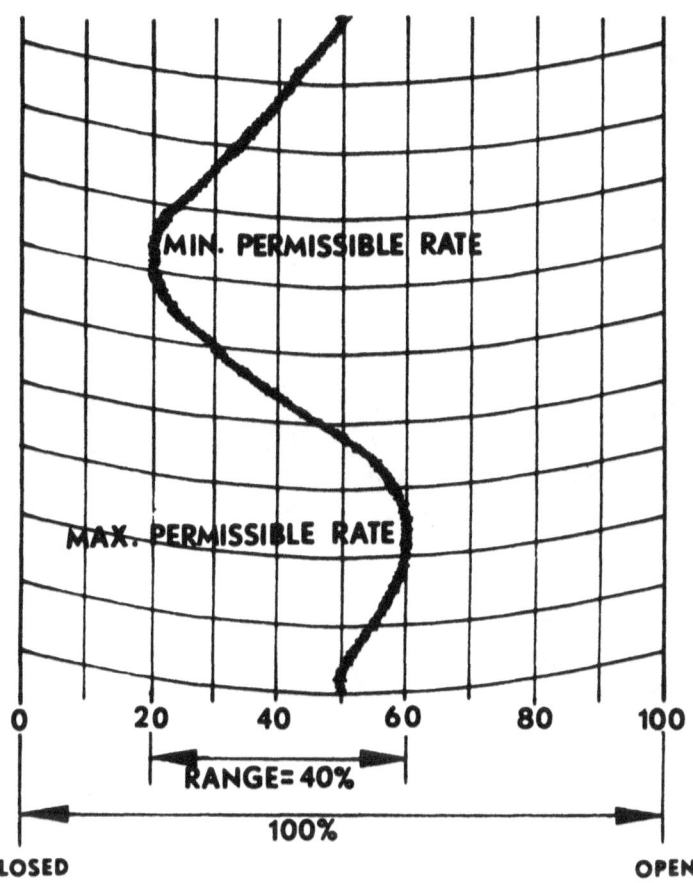

Fig. 15.4   Range on this output variable (fuel rate) chart, being 40% of maximum range, reaches from the 20% to the 60% lines of the chart.

in a tank or bin, if the instrument that measures this level does not work or measure properly? The answer to this question is obvious and needs no further clarification.

In order that this discussion is not drawn into fields and details beyond the control of the operators, assume that all other possible obstructions and important factors for the installation of an automatic control system have been taken care of and all necessary equipment for this type of control has been installed.

Fig. 15.5   Each of these flow transmitters sends measurement signals to a recorder or controller. (*The Foxboro Co.*)

15.2 TUNING (PROGRAMMING)

The first and most important step is the *tuning* of the controller. Tuning simply means making the controller do what we want it to do, by setting a set point and tuning certain dials to get an optimum response from the controller. When these settings are properly made, the controller will make the correct response to signals received from the process equipment. When these settings are incorrectly adjusted, regardless of how well and how elaborately the system has been designed, the end results in control will not be satisfactory. An automatic controller or a computer is

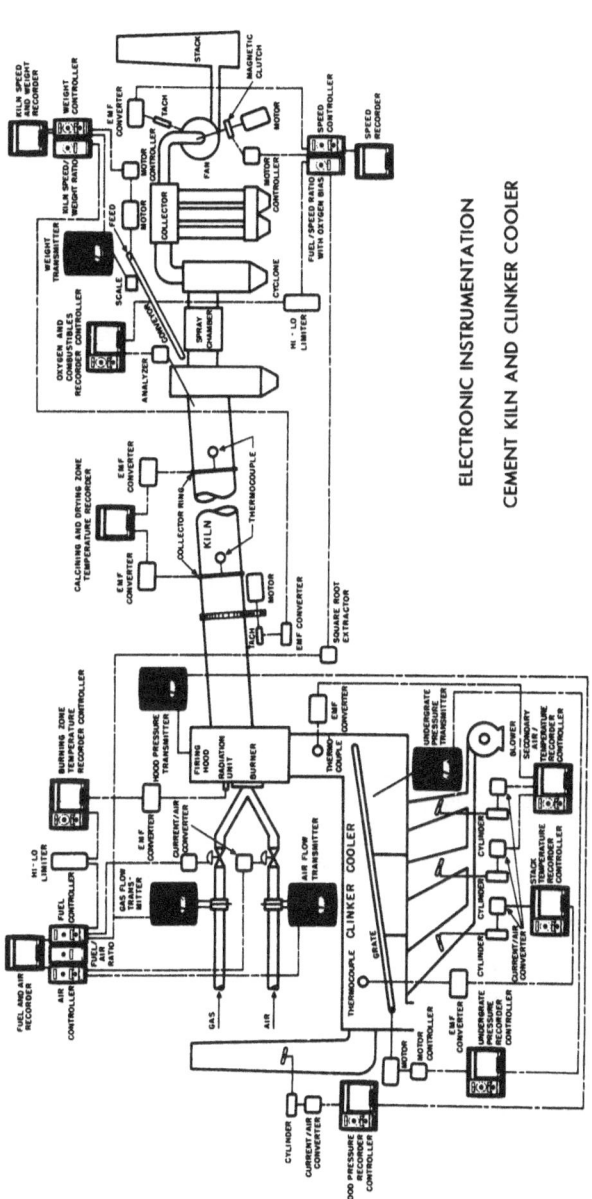

Fig. 15.6 Typical instrumentation for kiln and cooler. (*The Foxboro Co.*)

capable of doing only what it has been instructed to do—nothing more and nothing less. A mistake on automatic control is far more serious than a mistake on a process that is manually controlled. While an operator in manual control might make a mistake once in a while, an automatic controller will repeat the same mistake time after time without end, if programmed improperly.

In order to arrive at the proper setting for controller action on any process loop that is to be controlled, there are several variables that must be known before the control can be established.

First is the *span*, which is the range that can be tolerated between low and high values of the input variable. It is expressed as a percentage of the maximum span. Second is the *maximum span*, which is the largest possible variation of the input variable that can be measured on the instruments. Maximum span is normally considered to be 100% of the chart. These relationships are illustrated in Fig. 15.7. Third, the *operating range* is the desired percent range between minimum and maximum allowable limits in the value of the output variable (Fig. 15.4). Usually it is not possible to operate the output device over its entire range (e.g., a valve from fully open to fully closed) so an operating range must be selected that will provide the most efficient control.

In any process-control system, the *setpoint* is the ideal value of the input variable about which the process is controlled. Because many recorder charts are graduated in percent of chart,* it is customary to locate the setpoint at a point expressed by a certain percentage of the chart, usually 50–70%. For example, consider a linear recorder so designed and operated that its maximum span of 100% represents a temperature range from 2000° at zero percentage of chart to 3200° at 100% of chart. Maximum span of this chart (0–100%) represents a range of 1200°. Assuming that 2600° is the ideal temperature at which a good clinker can be produced without overburning or underburning, then 50% would be selected as the setpoint. Span would then be selected to cover whatever extent is allowable for the temperature to vary. Fig. 15.8 shows the relationship between percentage points on the chart and actual values being charted, in this case temperature. The operator in Fig. 15.9 is manually adjusting the setpoint on a recorder.

A convenient conversion table can be developed for the fuel rate (see Table 15.1). By using this table, the operator can determine the quantity

---

* The trend now is towards charts graduated in actual engineering units rather than in percent.

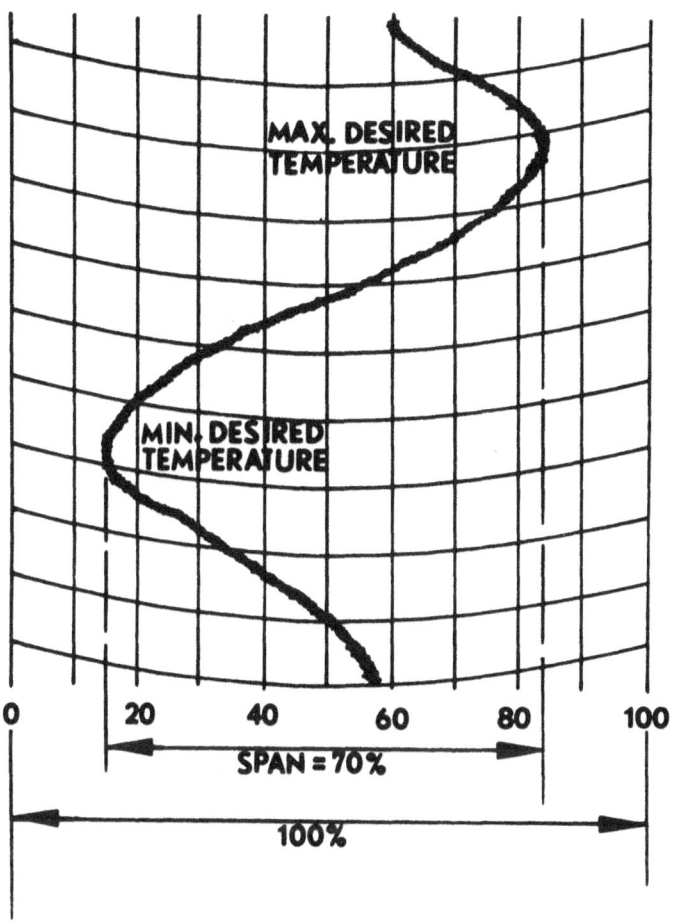

Fig. 15.7 Relationship of span and maximum span on a typical input variable chart for burning-zone temperature for which span, reaching from the 15% line to the 85% line, is 70% of maximum span.

of heat input at any fuel flow rate for either coal, gas, or oil at any percentage point on the recorder chart. This information is useful in determining whether the fuel is being used efficiently at any certain clinker production rate. It also tells the operator what equivalent setting to use when changing from one fuel to another.

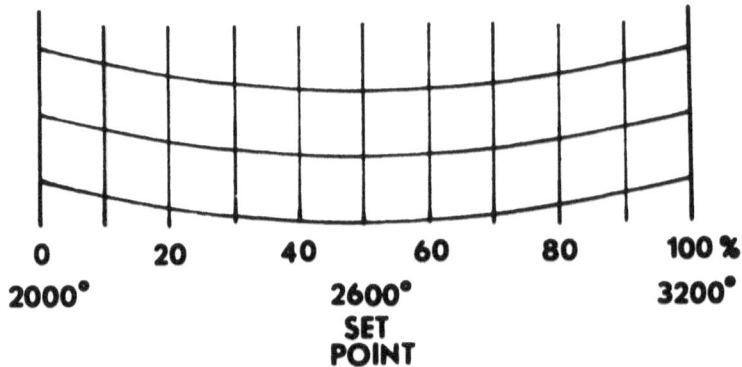

Fig. 15.8 Relationship between percentage on the chart and actual values, in this case temperature, is shown.

TABLE 15.1

FUEL RATE CONVERSION*

Gas rate				Oil rate			
Percent chart	ft³/h	MBtu/h	BBl clinker/h	Percent chart	gal/h	MBtu/h	BBl clinker/h
3	69.6	74.9	96	20	480	72.5	93
12	138.8	149.5	192	42	1008	152.2	195
15	155.2	167.2	214	46	1104	166.7	214
20	179.2	193.0	247	53	1272	192.1	246
26	204.0	219.7	282	61	1464	221.1	283
36	240.0	258.5	331	71	1704	257.3	330
40	253.2	272.7	350	75	1800	271.8	348
100	400.0	430.8	552	100	2400	362.4	465

* This table was extracted from a larger table for certain specific fuels and equipment. Similar tables can be computed for any plant.

In the discussion of manual control an example was cited in which the factors to be considered were described by a kiln operator before and after an adjustment in fuel rate was carried out. For automatic control to be

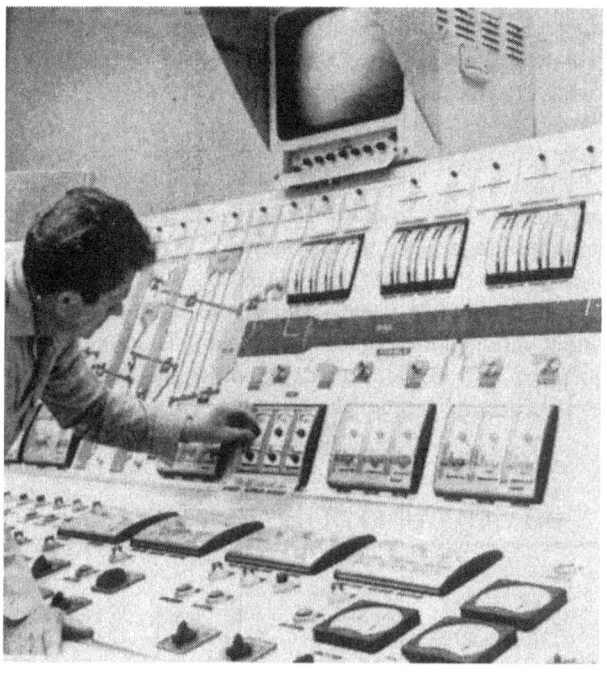

Fig. 15.9 The operator is adjusting the setpoint dial. (*The Foxboro Co.*)

successful, the controller must be programmed in such a way that it controls the temperature and adjusts the fuel rate in a similar fashion.

Most controllers have either one, two, or three action controls by which this can be accomplished. They are: *proportional, reset,* and *rate* action. In most applications proportional plus reset functions (two-mode) are utilized. When rate is utilized as a compensation for time lag in measurement, it is generally in a controller with proportional, reset, and rate function (three-mode).

Proportional Action. In a proportional-action controller, the output variable, or control means, is set in a specific relationship to the input variable, or process variable. The mechanics of setting the controller is simply an adjustment of a dial on the controller that shows the respective settings. A large range of settings is available depending on the particular instrument in use.

An important concept involved in proportional-action control is the *proportional band*, this being the ratio of desired span (input) to operating range (output) expressed as a percentage. That is

$$\text{Proportional band}^* = 100 \; \frac{\text{Span}}{\text{Operating range}} \qquad (15\text{-}1)$$

If the output variable is allowed to move through an operating range of 100% then the proportional band equals the span. If, however, the output operating range is less than 100, then the proportional band varies, as shown in Table 15.2 by some values selected at random. At a constant span for the input variable, the proportional band is increased to decrease the output range. Obviously, identical proportional bands are possible under completely different conditions in process control.

TABLE 15.2

VARIATION IN PROPORTIONAL BAND

Example number	Input variable Burning-zone temperature percent chart			Output variable Fuel rate percent chart			Percent proportional band
	Max.	Min.	Span	Max.	Min.	Range	
1	95	25	70	28	24	4	1750
2	90	30	60	30	22	8	750
3	80	35	45	30	24	6	750
4	85	35	50	32	20	12	420
5	83	38	45	34	19	15	300
6	80	40	40	35	18	17	235
7	75	45	30	38	16	22	136
8	90	30	60	56	12	44	136
9	70	50	20	40	14	26	77
10	67	52	15	41	13	28	54
11	80	20	60	100	0	100	60

Again referring to the example in which burning-zone temperature is being controlled, Fig. 15.7 illustrates the concept of span and maximum span for the input variable under control, in this case burning-zone temperature. Span, arbitrarily set at 70% of maximum span, is selected as

---

* Note that "gain" is the inverse of proportional band.

being the range of temperature that can be controlled by fuel-rate adjustments alone. In other words, the temperature will be permitted to vary over this range. Whenever the temperature starts to vary from the setpoint, the control is actuated and the fuel valve opens or closes as required depending on whether the temperature is below or above the setpoint. This is illustrated in Fig. 15.4 which shows the fuel-valve action over an operating range of 40% in response to the temperature input shown in Fig. 15.7.

Selection of proportional band settings can have a significant effect on the reaction of the output device. Fig. 15.10 is a stylized drawing of a recorder chart showing the burning-zone temperature deviating from setpoint. The output variable response is large or small depending on the proportional band. The relationship can be computed by the equation:

$$m = \frac{100e}{p} \tag{15-2}$$

in which

m = percent chart response of output device
e = percent chart change in input variable
p = proportional band setting.

Example: What percentage will the fuel rate change on the chart when the burning-zone temperature changes 20% on the chart, at a controller proportional band setting of 300?

$$m = \frac{100 \times 20}{300} = 6.7$$

Answer: The fuel rate will change 6.7% on the chart.

Proportional bands are changed according to the response of the process. For example, a water-level recorder might be set once and never changed. Kiln burning-zone temperature control, on the other hand, is more dynamic, and requires daily checking by an experienced person who is qualified to make any adjustments that may become necessary. Proper evaluation will avoid overcontrol, a common failing of the inexperienced operator, who is apt to make two or more simultaneous changes on a controller. Changes should be made one at a time, each change evaluated before going to the next one (if another one is actually necessary).

The serious shortcoming of proportional action control is the failure of the system to control at the setpoint in the process under control. This

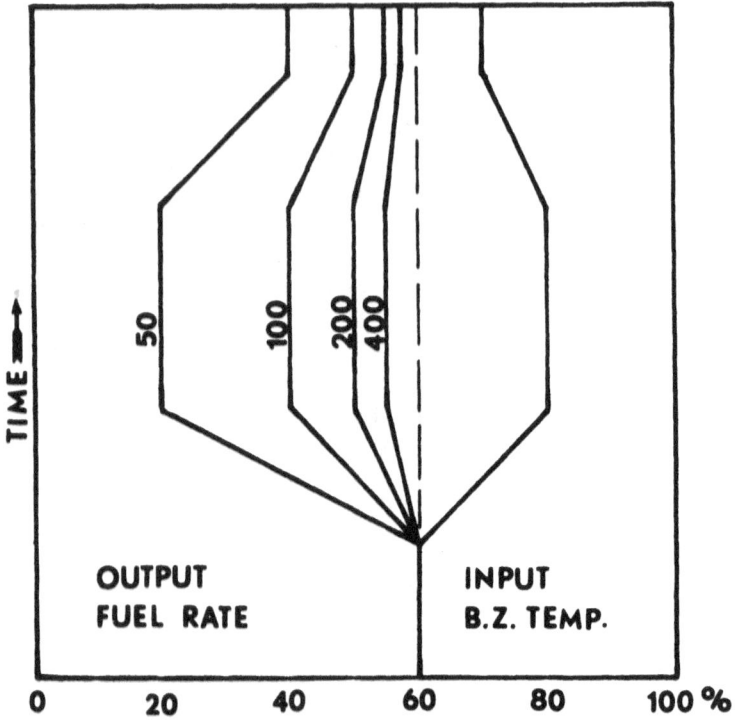

Fig. 15.10 PROPORTIONAL ACTION. Fuel-valve response to a change in burning-zone temperature at various proportional band settings.

means that proportional action does not necessarily bring the process variable (input variable) back to the desired setpoint when it deviates. All it does is to move the output variable in proportion to a change in the input variable.

Reset Action. Most processes require that the process variable be held at or returned to a specific setpoint for optimum and efficient operation. We have learned that proportional action does not consider the setpoint and will not necessarily return the process variable exactly to the setpoint once the process has had an upset. Another aspect of process control unaccounted for in proportional action is the process reaction time or lag. Often several minutes can pass after an adjustment has been made to the output variable before the process reacts to the change. With reset action, the means are available to return the process back to the setpoint and to account for possible delayed process reactions.

Reset action causes the output device to change at a rate proportional to the deviation of the input variable from the setpoint. How often this action is carried out (repeated) is governed by the *reset-time* setting on the controller. Reset time can be expressed as minutes per repeat or seconds per repeat. For example with a reset-time setting of one minute the action is repeated once every minute. With a setting of 22 min, the reset action would be repeated once every 22 min. Reset action enters into the control as long as the process variable is deviating from the setpoint. As the departure of the input variable from the setpoint decreases, the amount the output device corrects is decreased, during the course of any correction. When the process variable levels off at the setpoint, reset action will stop and does not enter the process again until the variable starts to move away from the setpoint.

This holds true in our example of burning-zone temperature control. When reset action is applied, the percentage change in fuel feed (fuel-valve movement) is identical to the percentage deviation of the burning-zone temperature from the setpoint. Added weight must be given to the setpoint in this instance because too high or too low a temperature can not only damage the quality of the clinker, but can also damage the coating and refractory as well as impair the overall operation of the kiln itself. With a change in the fuel rate the burning-zone temperature does not return to an acceptable level immediately thereafter. As pointed out previously, several minutes can pass before the temperature starts to reverese its deviation and move towards the setpoint. This delayed reaction can be observed in other areas of a rotary-kiln system. For example, when the bed-grate drive speed in the clinker cooler is increased in order to reduce the bed thickness, a considerable length of time can pass until the thickness is at the desired level.

The manner in which changes in reset time affect the process is shown in Fig. 15.11, which demonstrates that the output variable or control means (fuel response in this example) can be delayed or accelerated in response to departure of the input variable (burning-zone temperature) from the setpoint. Reset times (in minutes) are indicated by the figures adjacent to the fuel-rate response lines.

Reset rate, sometimes used instead of reset time, is merely the reciprocal of reset time. For example, if reset time is 2, then reset rate is 0.5 repeats per second.

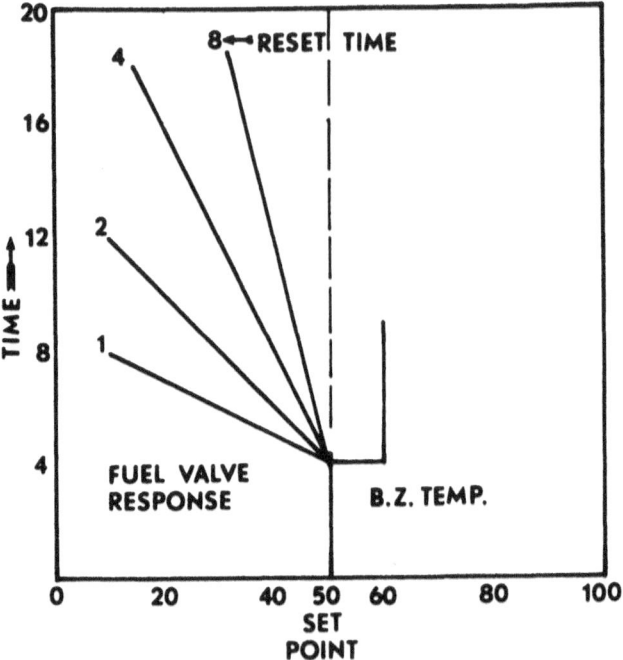

Fig. 15.11   RESET ACTION. A long reset time results in smaller output response compared with a short reset time.

Reset action response is practically instantaneous whenever the temperature moves away from the setpoint, i.e., it reacts immediately to the temperature difference or error and starts the correction. This response of reset action has been mathematically expressed.

$$\text{Percentage movement of output variable} = \frac{1}{\text{reset time}} \times \text{percentage deviation of input variable} \qquad (15\text{-}3)$$

Thus, when the temperature suddenly changes 15% and the reset time is 5 min, the fuel-valve response would then be:

$$\frac{1}{5} \times 15 = 3.0 \text{ percentage points per minute}$$

Converting into response per second yields 3.0/60 = 0.05 percentage points per second. Another example: The temperature changes 15% but this time the reset time is 20 min:

$$\frac{1}{20} \times 15 = 0.75 \text{ percentage points per minute or}$$
$$0.0125 \text{ percentage points per second}$$

From this formula another important equation can be established:

$$\text{Reset time} = \frac{\text{percentage deviation of input variable from setpoint}}{\text{percentage movement of output variable in a given time}}$$

Remember to use the same time units in the result as stated in the output variable movement. That is, if the output variable movement is expressed in seconds, then reset time is likewise expressed in seconds.

Reset action eliminates cycling conditions in the process caused by the earlier-mentioned reaction time after an adjustment has been made to the control device. In setting reset time on the controller, one usually starts with a long reset time and progressively reduces the time setting until the cycling condition stops.

Derivative Action. The third and last control action available on some controllers is derivative action, or *rate action.* Derivative action is usually combined with proportional, or proportional plus reset action, and is almost never used alone to control a process. It is used as a compensation for time lag or inertia of the measured variable.

Derivative action causes the output device to respond proportionally to the *rate* of change of the input variable when it is changing. In our example, when the burning-zone temperature moves rapidly in either direction, the fuel-rate response will be correspondingly fast, and when the temperature moves slowly, the fuel-rate response will be correspondingly slow. In essence, derivative action is the same as the procedure most kiln operators commonly use when they control a rotary kiln manually. When they look into the burning zone and notice a large, rapid temperature change, they "hit" the kiln with a large change in fuel rate. When the temperature changes only slowly, they "nurse" the kiln with small adjustments in the fuel rate. Derivative action diminishes to zero when the burning-zone temperature stabilizes. One must remember that derivative action, like proportional action, does not consider the setpoint. In other

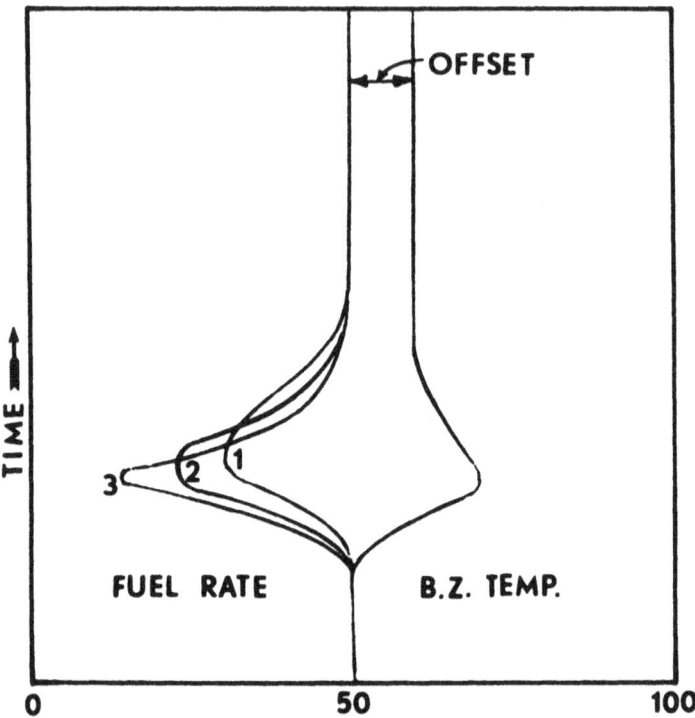

Fig. 15.12 DERIVATIVE ACTION. In Curve 1, the derivative time (fuel-rate response) is relatively slow, in response to a slow rate of change of the input variable (burning-zone temperature); Curves 2 and 3 represent progressively shorter derivative times.

words, derivative action ceases as soon as there is no more change in the input variable, e.g., the burning-zone temperature, and does not consider whether the temperature levels off at the setpoint or at some other temperature.

Derivative action is adjusted on the controller as a function of time just as in reset action. Thus, one refers to *derivative time* whenever an adjustment has to be made on the controller. In Fig. 15.12, Curve 1 represents a long derivative time (slow response), and Curve 3 represents a short derivative time (fast response).

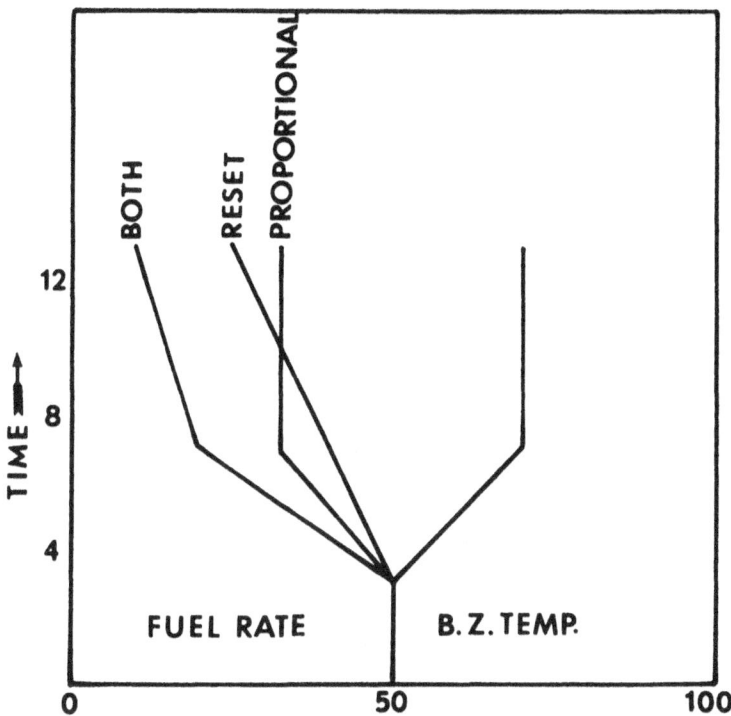

Fig. 15.13 Combined proportional plus reset action is shown.

Proportional Plus Reset Action. From the previous discussion, it can be seen that proportional action sets the output device proportional to the change in the input variable, and reset action changes the output device at a rate proportional to the deviation of the input variable from the setpoint.

Fig. 15.13 shows how these two actions can be combined, again using burning-zone control as an example. When no reset action exists (reset time set at infinity), the fuel-rate response would be identical to the proportional action. However, with the two control methods combined, the fuel rate is influenced by the reset action. Thus, at time increment 7 the temperature has leveled off, consequently proportional action has stopped because temperature is no longer changing.

Reset action, however, keeps the fuel rate decreasing because the process is off setpoint, and ceases to enter the process only when the process has returned to setpoint. With the two control methods combined, reset time is the time required to repeat the proportional response of the controller.

The response of the output device, under the two combined control actions, can be computed by the equation

$$m = \frac{100}{p} \left(\frac{1}{t} + 1 \right) e \qquad (15\text{-}4)$$

in which:

   m = percent chart response of output device
   e = percent chart response in input variable
   p = proportional-band setting
   t = reset-time setting.

Example: What percentage will the fuel rate change on the chart every second when the burning-zone temperature changes 20% on the chart and the controller settings are: proportionall band, 300; reset time, 5 s.

$$m = \frac{100}{300} \left(\frac{1}{5} + 1 \right) 20 = 7.99$$

Answer: The fuel rate will change 8% on the chart each second.

Application. At a first glance, it appears that not many tools are available with these three control actions to control a complicated process as described in the examples. However, this is not so, because the possibilities to vary these settings are almost unlimited. Table 15.3 summarizes the three basic control systems, and Fig. 15.14 illustrates how the output variable reacts under each of the systems.

Often, time and personnel to study the process beforehand are not available, which makes it necessary to find the proper settings on the controller by trial and error. In many instances, this is easier to do than calculating the settings. Many controller manufacturers recommend a wide proportional band setting. Working first by reducing the proportional band, then increasing the derivative time or reducing the reset time, an attempt should be made to eliminate process cycling. This procedure is repeated until the minimum reset, minimum proportional band, and maximum derivative time can be found so that cycling no longer exists. These settings give maximum response compatible with stability of control.

In the foregoing discussion, it has been shown how proportional, reset, or derivative action can automatically control a given process in the kiln

TABLE 15.3

SUMMARY OF CONTROLS

Control	Action	Example	Remarks
Proportional	Output device moves in a manner directly proportional to a change in the output variable.	Fuel valve is opened (or closed) an amount directly proportional to the drop (or rise) in temperature in the burning zone.	Large response requires small proportional band; small response requires large proportional band.
Reset	Output device moves in a manner proportional to the departure of the input variable from the set-point.	Fuel valve is opened (or closed) an amount proportional to the amount the burning-zone temperature is below (or above) the set-point.	Large response requires short reset time; small response requires long reset time.
Derivative	Output device moves in a manner proportional to the rate of change of the input variable.	Fuel valve is opened (or closed) an amount proportional to the rate the burning-zone temperature is falling (or rising).	Large response requires a long derivative time; small response requires short derivative time.

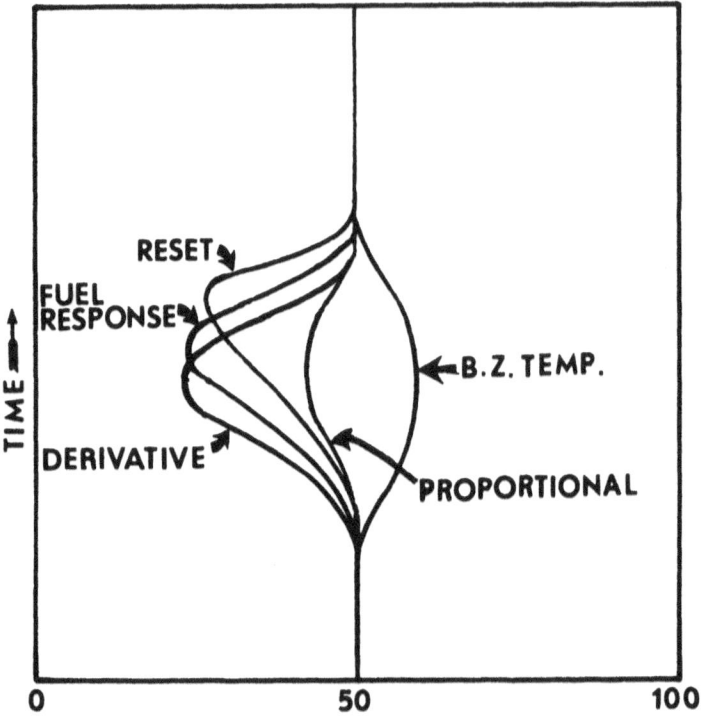

Fig. 15.14   A typical summary of proportional, reset, and derivative action combined on one chart.

system. It is also apparent that programming a controller is considerably more complicated than manual control. Finding the correct setting is a time-consuming task requiring preliminary process studies and process knowledge before the correct answer can be found. One should always keep in mind that the rewards from serious efforts made in this direction can be large not only in clinker quality and operating efficiency but also in the ease of operation. Without any doubt, one thing that gives the kiln operator the most satisfaction on the job, is to have the kiln operation so stable that only one or two small adjustments must be made during the entire shift.

15.3 KILN-CONTROL LOOPS

As previously mentioned, kiln-control concepts have become quite elaborate and sophisticated. Automation started with simple individual control loops that involved one input to a controller which in turn sent a signal to the output device to turn a motor-activated valve, damper, or drive. Technology makes it possible today to transmit and to process hundreds of process-variable inputs simultaneously to a process computer. This microprocessor, in turn, might be programmed and made capable of making a control-adjustment decision for one single output based on simultaneous analysis of ten to twenty different pieces of input data.

Single Closed-Loop Control. In single closed-loop control, a measurement of the variable (such as temperature, pressure, etc.) to be controlled is fed to the controller (or computer) which compares it with a setpoint and makes any adjustment in the output if there exists an error or deviation from this setpoint.

Some of the more common loops in this class are:

SINGLE CLOSED-LOOP CONTROL

Input	Output
Fan inlet total pressure	→ Fan damper position
Hood pressure	→ Cooler stack damper position
Undergrate pressure	→ Cooler grate drive speed
Burning-zone temperature	→ Fuel rate
Coal mill outlet temperature	→ Mill inlet ambient air damper
Cyclone outlet temperature	→ Cyclone inlet ambient damper
I.D. fan inlet temperature	→ Water spray rate or ambient damper
Kiln speed	→ Feed rate

INTERLOCK LOOPS

Input	Output
I.D. fan down	→ Fuel valve inoperative
Primary air fan down	→ Fuel valve inoperative
Gas analyzer (CO for > 2 min)	→ Power to electro prec. off
Coal mill outlet temp. above limit	→ CO_2 to coal mill on

Cascade (Computer) Control.   This is the application area that is perfectly suited for computers because in cascade control several process variables are evaluated before a decision about a necessary change is made. Take for example the aforementioned *burning-zone temperature—fuel rate,* closed-loop control concept. In this single loop the fuel rate is adjusted based strictly on the deviation from the setpoint of the burning-zone temperature, regardless of the oxygen content in the kiln gas or the reasons for the drop in temperature. Kiln operators would agree that this would be a very primitive way of controlling the burning-zone temperature because they themselves would *not* take such a simplistic approach when they control the kiln manually. For example, the operator's thought process would follow this path:

1. Notice that burning-zone temperature is dropping
2. *Questions:*   a) current and previous percent oxygen in exit gas?
   b) current and previous kiln-torque or kiln-drive amps?
   c) current and previous kiln exit-gas temperature?
   d) current and previous I.D. fan speed?
   e) current and previous feed loading in kiln?
   f) current and previous kiln speed?
   g) current and previous fuel rate?
3. *Decision:* *If* a to g = this or that, *then* adjust . . .

The change in burning-zone temperature is the result of a disturbance that exists or happened earlier in one or several of the above-listed variables. In cascade control, this mental work is being done in a split second by a computer which evaluates these different variables, looks at their trends, and makes the adjustment based on a program that is stored in its memory. Advanced computer programs are capable of predicting the change in burning-zone temperature before it actually takes place or becomes noticeable to the operator. Coal-grinding and cooler circuits are other areas of the kiln system where this control concept has found wide acceptance. Complete automatic-control programs and systems are in existence today that can take a kiln from a cold start, control it for an indefinite length of time, and take it to a complete shutdown, without the operator having to turn a knob. Such modern control systems usually also incorporate graphic flow diagrams and trend displays of any circuit or variable the control-room operator chooses. Most important of all is that all these systems have manual control back-up capabilities to allow an operator to take over with manual control when the computer is down for maintenance. Hence, regardless of how sophisticated or automated a kiln-control system is, there is clearly a need to train control-room operators in manual control of the kiln.

One can also discount the fear that, as the evolution of computer control will continue, there ultimately will be no more need for kiln operators. Perhaps it is true that the operator will have very little work to do in the future, however, there always will be a need for this skilled position. The operator's job will become easier and the mental stress factor, that was so prevalent in the "old" times of complete manual control, will undoubtedly be greatly reduced in the foreseeable future.

16.

Kiln Control Variables

16.1 BURNING ZONE

The burning-zone temperature is not only one of the most important kiln control variables but also the most difficult one to monitor. Despite the fact that burning-zone conditions in modern kilns are measured and monitored by means of sophisticated instruments, kiln operators should always be thoroughly trained to visually observe and evaluate the burning zone. This capability for visual inspection is an absolute requirement because there are times when these instruments are out of service. Visual verification is also needed when instrument readings are questionable due to dust interference in the burning-zone environment.

This control function is not merely a check to see whether good clinker is being burned, but is also an evaluation of the clinker, the coating, the flame, and the air stream. Color is important in evaluating the clinker, the coating, and the flame. General behavior of the clinker, air stream, and flame can be studied, as well as shape of flame and coating, and size of clinker being discharged.

The pyrometer records the temperature of a comparatively small area. A change in temperature registered by this instrument does not always mean a true overall temperature change in the burning zone, and therefore does not always indicate a need for adjustment in the kiln controls. As a matter of fact, a change in pyrometer reading could be caused by a shifting of the burning zone due to a change in the flame characteristics, or by dust interference entering the burning zone with the combustion air from the cooler.

With reference to the television monitor, it must be kept in mind that the camera does not "see" the burning zone as a whole. What can be seen on the TV screen is only a fraction of the zone. Here again, dust interference could make visibility in the burning zone poor or even impossible at times. Finally, either one or both instruments could fail. All these factors emphasize the need for training a new operator thoroughly in the art of "eye-balling" the burning zone.

Viewing the Kiln Interior. At this point it is important to learn how to look properly into the burning zone. Although this seems elementary, it should not be forgotten that the burning zone is an extremely luminous light source. Although filtering glasses are used, the light source is so strong that focusing the eyes into it for too long a time could cause partial blindness. One should look no longer than one minute at a time into the fire. If longer viewing is required, look aside for a few seconds occasionally to rest the eyes. Looking steadily too long at the flame results in the eye losing its ability to see details.

The question of what type of colored filter glass to use must be left to each individual operator, as one person can see better with one particular glass than with another, but the same glass may not suit someone else. Burning with a natural gas flame usually requires a darker colored glass than an oil or coal fire would, because of the greater luminosity of a gas flame. As a rule, one should always use a glass that enables him to see under and behind the flame. Once a certain glass has been chosen, the operator should stay with this glass at all times in order that proper judgment of the conditions in the burning zone can be made. Sometimes an operator has the habit of overburning the clinker, and another may have the reputation of consistently underburning the clinker. A very effective measure to counteract these habits is to equip the "hot" operator with a brighter glass and the "cold" operator with a darker glass to compensate for their misjudgment of the burning-zone temperature.

Many control rooms, in modern cement plant design, are located far away from the firing floor and thus demand more reliable and accurate instrumentation to monitor the burning-zone temperature. Some of these plants are highly successful in this endeavor but there are also plants where this is a source of concern. Generally, it is easier to satisfy the requirement of recording accurate burning-zone temperature trends on kilns with little dust interference. On the other hand, dry-process kilns, with their commonly poor visibility, are much more difficult to monitor for burning-zone temperature. Although desirable, absolute accuracy as to specifying prevailing temperatures is not so important as the true temperature trend that takes place in the kiln. In other words, as long as the instrument registers a corresponding temperature drop when the kiln is actually cooling down and an increasing temperature when the kiln heats up, the requirement of temperature monitoring of the burning zone is fulfilled.

How frequently should one look into the burning zone? There is no set answer to this question. Naturally, on remote-controlled kilns there is no need for the operator to pace the firing floor as frequently as on kilns that are almost exclusively manually controlled. It is on these manually controlled kilns where experienced operators sometimes become overconfident and think that it would be perfectly safe to leave the kiln alone for periods in excess of 40 min. This action, however, is against good burning practice. The secret of every good operator is his ability to recognize a change in kiln condition at the time a change takes place and not later. For this reason a good operator will never leave a kiln unchecked for too long a time. When things are going smoothly, the kiln should be checked every half hour, with more frequent checks if adjustments are being made.

Appearance of Clinker. The quickest, although not the most accurate, means to check the clinker for quality is by observation of the color and the size of the clinker. Some operators have become so proficient that they can tell fairly well how good the clinker has been burned by merely looking at a handful of clinker. It should be pointed out, however, that clinker size alone does not give a true indication of clinker quality as clinker size is influenced by such factors as feed composition and kiln speed. Generally speaking, well-burned clinker is dark, almost black in color, and the "hotter" the clinker is burned, the larger the clinker nodules become.

If the clinker is overburned (that is, burned hotter), the free-lime content drops, the liter weight gets heavier, the clinker gets larger in size, more dense (less porous), and darker in color, compared with a clinker that is not overburned. Conversely, underburning the clinker (burning colder) causes the free-lime content to increase, and the liter weight to lighten. Also the clinker gets smaller in size (dustier), more porous, and the color is lighter—more nearly brown.

Until now the appearance of the clinker after it has been burned and cooled has been discussed. Now consider the clinker in the burning zone. For this a reliance on eyesight is necessary.

The kiln feed, when it approaches clinkering temperatures, undergoes drastic changes in its physical and chemical characteristics. As soon as liquid formation of the constituents begins, the feed becomes sluggish because the chemical reactions tend to make the feed soft and viscous. Formation of balls of feed (clinkers) now commences, and the material starts to ride up higher on the rising wall of the kiln. Instead of sliding down the wall, the feed bed starts to cascade over itself. By noting this cascading action in the vicinity of the flame the observer will notice an increase in sluggishness of the feed at higher clinkering temperatures, the feed will climb higher on the kiln wall, with more turbulent cascading. Assuming that other variables remain constant, larger clinker balls are formed at higher temperatures.

Now consider the color of the clinker. Any color seen in the burning zone can be directly related to the temperature. Because different filter glasses are used for viewing, it is difficult to assign a definite temperature to any certain color. In very general terms, however, the kiln condition can be estimated from the colors observed in the hottest part of the flame:

dark red	cherry red	orange-yellow	white
cold		normal	hot

Any deviation of color from the orange-yellow should be investigated to determine the cause and what adjustments need to be made to the kiln operation.

The Feed Behind the Flame. The manner in which the operator controls the position of the raw feed has a significant influence on operation of the kiln, and is one of the important key indicators of operating stability. Changes in this raw-feed posiiton are an early warning signal that burning-zone conditions are about to change.

Special attention has to be given to the physical characteristics of the feed behind the flame. Although in most cases it will be difficult for the operator to see behind the flame because of the flame shape and dust interference, nevertheless, no effort should be spared in trying to see as far back as possible, because it is in this region of the kiln where an early detection of possible kiln upsets can be made.

The Dark Feed. When one looks into the burning zone, one will observe a sharp color change of the lower part of the feed bed under the flame from dark to bright, as shown in Fig. 16.1. This point in the burning zone is of great importance to the operator. The importance of this lies in the fact that the position of this *dark feed* gives the kiln operator one of the earliest indications of when the burning zone tends to warm up or cool down. Under normal and stable conditions the position of the dark feed remains stationary approximately one-quarter of the distance into the flame (as shown in the figure). If it moves farther under the flame (towards the front of the kiln), the burning zone is cooling down; if it shifts in the direction of the kiln rear, the burning zone is warming up.

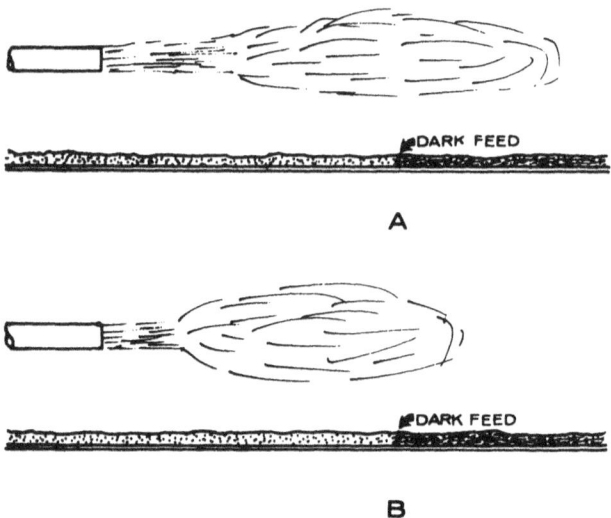

Fig. 16.1 The dark feed should be about one-quarter of the way under the flame.   If the dark feed slips too far under the flame, as at A, its relative position can sometimes be changed by adjusting the flame length, as at B.

Any change in flame characteristics will have a direct effect on the position of the dark feed. For example, whenever the flame is shortened the dark feed will move closer to the discharge end of the kiln. A change in position of the dark feed therefore does not always mean that the burning zone is either heating up or cooling down. The position of the dark feed can move because of a change in the shape of the flame, a change in burnability of the feed, or a change in the feed loading of the kiln. The dark feed will move closer (under the flame) when the flame is shortened, the feed is harder to burn, the feed loading of the burning zone increases, or the burning zone cools down. It will move away from the flame (toward the feed end) when the flame is made longer (and enough heat is available to drive the dark feed back), the feed is easier to burn, the feed loading in the burning zone decreases, or the burning zone heats up.

Any change in the position of the dark feed must be viewed in the light of all of these influences. The operator must be able to see the dark feed whenever he looks into the kiln, and he must regulate operation of the kiln so as to achieve this end. A slow shifting of the dark feed in either direction can usually be counteracted by a small change in the fuel input rate in order to keep the feed in its proper place.

The dark feed should never be allowed to move further under the flame than one-half of the flame length. When it becomes necessary to counteract the condition in which the feed has "slipped" too far under the flame, the operator can, if the means to do so are available, change the flame length to restore the dark feed position to the proper relation with the flame again. Fig. 16.1 shows how this is accomplished. Figure A depicts a condition in which the dark feed has advanced too far under the flame. By increasing the primary air pressure and temperature, and if possible the secondary air temperature, the flame will be shortened as shown in Figure B. Note that the position of the dark feed has not changed, but the relation between feed and flame is in its proper perspective again because the flame is shorter. This shortening of the flame results in more heat being released in the critical area where it is most needed.

Some operators take advantage of this procedure quite freely whenever they encounter the described condition. However, one has to consider another aspect caused by such a change in the flame structure. The shorter fire results in a bushier flame that is more in contact with the coating, the coating being thus exposed to a greater heat. For this reason a shortening of the flame should never be carried out before the operator has assured himself that the coating will be able to withstand the extra heat. For example, if a hot spot on the kiln shell is already in existence in the critical area, it would be a serious mistake to shorten the flame and release more heat over the already weakened area.

Combustion Air From the Cooler. Considering only the general appearance of the air coming from the cooler under normal operating conditions, not much attention need be given to this factor because the appearance remains nearly constant. Under upset conditions, however, this factor becomes very significant.

When insufficiently burned clinker is allowed to enter the cooler, the airstream carries a large amount of fine dust particles back into the kiln which can obstruct visibility in the burning zone. Furthermore, the secondary air temperature is most likely low in such a case, causing a change in the flame structure, with the result that the ignition point of the fuel moves further into the kiln. Combustion conditions are poor (cold flame) because of the dust-entrained atmosphere in the burning zone. All these conditons may make the burning zone appear to the eye to be "cold," although this is not necessarily the case at all times. The area of highest intensity could already have approached normal clinkering temperature again although the front is still black and cold. Whenever dust obstructs the view in the burning zone, the operator must make a special effort to see under and behind the flame. Experience has established the rule that, whenever dusty conditions prevail in the burning zone, the clinkering conditions under and behind the flame, rather than the color of the front of the burning zone, govern subsequent corrective measures after an upset.

Color of the Coating. The color of the coating tells a great deal about temperature conditions in the burning zone. Under normal operating conditions, the color of the coating in the hottest area ranges between yellow and white. When the color changes to orange or red, the zone is cooling down; if it changes to white, the burning zone is heating up.

A large portion of heat is transferred to the feed by radiation and conduction from the kiln wall (see Chapter 2). Thus it becomes understandable that the temperature of the coating is very important for the burning process. In addition, the coating acts as a heat storage in the burning zone. Having a hot coating will in many instances enable the operator to fight out a heavy onrush of feed. On the other hand if the coating loses temperature rapidly at the time a heavy load of feed enters the burning zone,there is usually no alternative but to slow down the kiln to avoid bad clinker.

Besides the color of the coating, the general appearance of the coating is also important. The operator should try to detect weak spots, loss of coating, or formation of rings in the burning zone. Early detection is of utmost importance whenever a change in the coating structure takes place. It is the operator's duty to maintain or rebuild the coating to protect the refractory and the kiln shell from damage by overheating.

A condition that can cause a great deal of damage to the kiln if it is not corrected imediately is when the burning zone becomes so overheated that the clinker starts to ball up or even worse, begins to liquefy. This condition is extremely hard on the coating as the coating, becoming soft (liquefying), starts to come off. Whenever this happens, the operator has to disregard the production of good clinker, and concentrate on protection of the refractory and the kiln shell. Failure to make immediate corrections can result in red spots appearing on the kiln exterior, a sure indication that something is wrong.

In summary, it is important to visually inspect the burning zone for changes in the following items (any of which could signal the possibility of change in burning-zone temperature):

a) Clinker color
b) Clinker size
c) Cascading action of the clinker bed in vicinity of the flame
d) Feed-bed appearance behind the flame
e) Dark feed position
f) Appearance of secondary air coming from the cooler
g) Coating conditions
h) Flame shape and color

Knowing what to look for in the burning zone and being able to recognize changes in the items listed above doesn't necessarily indicate that one is now ready to control the burning-zone temperature. Simply stated and as mentioned earlier, burning a kiln is not merely a function of adding fuel when the burning zone gets cool or conversely, reducing fuel when the burning zone gets hot. Burning-zone temperature control must be considered with other kiln control variables before corrective action can be taken.

Since it is difficult to measure the actual burning-zone temperature with pyrometers, several (old and new) methods have been tried to relate other kiln variables with possible changes in this burning-zone temperature. The most recent and noteworthy method in this respect is the monitoring of the NO_x content in the kiln exit gases and relating this to the burning-zone temperature changes (See Fig. 16.2).

Kiln burning is a matter of detecting any pertinent changes as early as possible and making countering adjustments in small steps. It can best be described as a control by anticipation, i.e., trying to establish what the kiln conditions (burning-zone temperature) will be a few minutes later. Waiting to make an adjustment until the full effect of a large deviation is registered on the instrument can lead to unstable operation and is not consistent with good burning practice. This applies not only to burning-zone temperature but to all other main control functions.

How can an operator predict what the burning-zone temperature will be in, e.g., 10, 20, or 30 min hence? The proven indicators of approaching changes in burning-zone temperature are:

a) The kiln drive torque or amperage.
b) The oxygen content in the exit gases (provided the fuel rate and I.D. fan speed remained unchanged).
c) The back-end temperature (chain outlet, kiln inlet, calcining zone) at that time that equals the travel time of the feed from the

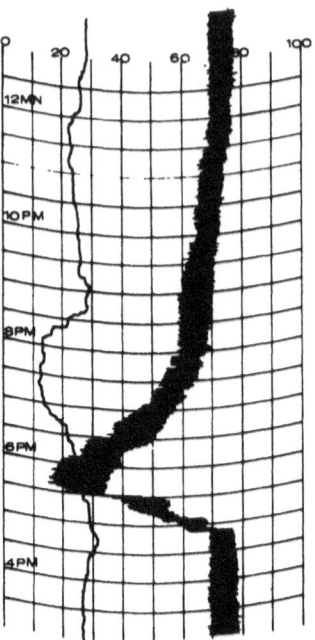

Fig. 16.2 The kiln slowdown was started in stages at 4:50 PM and completed at 5:30 PM. Then the speed was gradually increased again, as the kiln warmed up, and the kiln was back to full speed and balanced out at 12 AM.

respective thermocouple to the inlet of the burning zone.

d) The aforementioned NO_x content in the exit gases.

e) The CO_2 content in the exit gases.

f) The kiln-feed end draft.

These constitute the primary early-warning instrument readings that alert an operator to upcoming changes in burning-zone temperature.

16.2   KILN EXIT-GAS ANALYSIS

Oxygen, carbon monoxide (CO), and carbon dioxide (CO_2) have been extensively discussed in Chapter 5. From these discussions it has become apparent that the exit-gas analyzer combined with its O_2 and CO recorder in the control room is one instrument no kiln can do without. As a measure of safety precautions this is also one of the instruments that irrevocably MUST be in operation at all times while the kiln is operating, i.e., when fuel is fired into the kiln.

This gas analyzer, because it samples "dirty" kiln gases and takes the sample at a location where high temperatures prevail, has a tendency to malfunction frequently unless almost daily preventive maintenance is carried out on this unit. The location, where the sample probe is installed, is also a key point to consider as false air inleakage could distort the true contents of O_2 and CO_2 in the exit gases. Kiln operators must be on the look-out for such malfunctions and notify the foreman or instrument shop whenever a malfunction occurs or is suspected. Operating a kiln without the gas analyzer functioning properly is a dangerous thing to do and can lead to catastrophic accidents. This is one instrument that doesn't allow room for compromises; it simply has to work and must be repaired as quickly as possible when it doesn't. These are strong words but considered worth mentioning for the simple reason that many accidents have happened in the past in some cement plants because the gas analyzer wasn't working properly.

For example, a few years ago a kiln operator noticed an irregularity in the gas-analyzer recording wherein more than 2% carbon monoxide (CO) was indicated but, at the same time the oxygen content showed peak levels of 5%. In this example, the operator made the wrong assumption, specifically he believed the oxygen (O_2) to be correct and the CO recording to be in error. *Result:* A fire in the kiln back end. *Lesson:* Anytime that the presence of carbon monoxide (CO) is indicated, reduce the fuel rate immediately and ask questions later. In such situations it helps to re-member the triangular relationship for combustion discussed in Chapter 5. In the above example fuel was present (i.e., CO which is unburned car-bon), air for combustion became available from possible inleakage of ambient air into the system at the kiln back end, and the only missing link for a fire or explosion at the feed end was heat for ignition. The third component for this accident to occur was most likely provided by the high temperature of the exit gas or the electrostatic precipitator which by itself is an igniter in the true sense. To safeguard against such accidents, most electrostatic precipitators are interlocked with the fuel rate. In other words, the precipitator is automatically deenergized whenever CO is present for more than 1 min in the exit gas.

Earlier, it was mentioned that both too much (more than 3.5%) and too little (less than 0.7%) oxygen in the kiln exit gases represents inefficient operation and that 1.0–1.5% O_2 is a level that results in optimum operating conditions.   This applies to all kilns with the exception of a precalciner kiln that has no tertiary air supply to the flash calciner.   In these types of precalciner kilns, all the combustion air requirements for the flash calciner are supplied from the kiln, hence higher oxygen levels are required at the kiln exit to secure complete combustion of the fuel given to the flash furnace.   However, the same principle would apply at the preheater cyclone outlet, i.e., for the gases after the flash furnace.   Here too, the oxygen content should not be less than 0.7%, not more than 3.5%, and ideally between 1.0–1.5% during normal kiln operations.   In plants where low-grade fuels with large variations in heat value are used for firing in the flash furnace, it is advisable to hold the percent oxygen at a higher level than the indicated range to secure sufficient air availability when surges of higher heat-value fuels occur.   Since precalciner kilns operate with two combustion-process locations, it is advisable to equip these kilns with two gas analyzers, one at the kiln exit and the other at the flash calciner outlet.   In this manner, both combustion processes can be independently controlled from each other.   However, on precalciners that are equipped with only one analyzer at the preheater tower exit, an operator might have considerable difficulty in relating changes in O_2 or CO to either one of the two burning locations.

The question arises: "What is to be done when the kiln operates steadily but continuously with oxygen contents higher than say 3% or lower than 0.7%?"   In such instances, because of prevailing stable conditions, no drastic changes should be made as such moves could upset the delicate balance of the kiln.   Rather, it is advisable to fine-tune the controls in small steps, allowing ample time of about 1 h between each to make sure this balance and stability is not disturbed.   Fine tuning in this manner may take up to 8 or 16 h and is referred to as optimizing the kiln operation.   As with any other kiln control variable an operator follows the following basic steps in mattters of exit-gas control:

1. Secure the safety of the equipment and personnel (sufficient O_2 available and no CO showing)
2. Stabilize the kiln (O_2 to level off and stay within a certain narrow range)
3. Optimize (fine-tune) the O_2 in small steps to bring the oxygen content into the range of 1.00–1.5% to obtain optimum kiln efficiency.

The majority of kilns are equipped with analyzers that test only for contents of oxygen and carbon monoxide which are considered the two key variables in combustion control. Some kilns also have separate carbon dioxide analyzers and monitors (see Chapter 5) but these are not considered as absolute requirement, rather they are excellent tools to forewarn the operator of upcoming changes and are helpful in optimizing the kiln operation.

In recent years, a great deal of work has been done in some plants in the utilization of recordings of NO_x in the kiln exit gases for determining changes in burning-zone temperature. Originally these NO_x analyzers were installed and used for the purpose of emission control to satisfy environmental specifications. Engineers, being familiar with the process of NO_x formation in a rotary-kiln flame, have recognized the relationship between the NO_x in the exit gas and the burning-zone temperature. Different NO_x concentrations are found in the exit gas when the combustion temperature (flame temperature) in the burning zone changes. Higher flame temperatures result in higher NO_x. Furthermore, when the kiln excess air (O_2) increases, the NO_x also increases. It must be mentioned that NO_x cannot be related to the actual prevailing temperatures in the burning zone but it is an excellent and rapid indicator of changes that have taken place there. In many respects the above reactions to flame temperature and excess air in the kiln lead to contradictory objectives in kiln operation and have been the subject of much controversy. Naturally, in matters of environmental control, one strives to obtain a NO_x content as low as possible. For an efficient kiln operation, however, one tries to operate the kiln with as short a burning zone as possible (short flames, high flame temperatures) which in turn means high NO_x contents. This problem is especially acute on natural gas-fired kilns because they usually operate with higher NO_x contents than coal- or oil-fired kilns.

As a general rule, kiln exit-gas components are affected as follows, assuming all other variables remain constant.

Strip chart recordings of oxygen often display large variations in range even when the kiln operation is stable. This is referred too as process "noise" and is a normal occurrence on many kilns. However, when this so-called "noise" exceeds a range of ± 0.5% in short-time intervals, this could be an indication of irregular fuel or air-flow rates taking place within the system and should be investigated. Likewise, since process "noise" is common to the oxygen analyzer, the instrument should be checked whenever the oxygen chart traces a straight line for a long period of time.

Most oxygen-recording charts register this variable within the range of

		O_2	CO	CO_2	NO_x
Fuel	increases	down	up	down	down
	decreases	up	down	up	up
I.D. fan	increases	up	down	up	up
	decreases	down	up	down	down
Flame temperature	increases	=	=	=	up
	decreases	=	=	=	down
Feed rate	increases	down	=	up	=
(calcination)	decreases	up	=	down	=

0–5%. It is important to remember the following: When the recording is pegged out at the maximum for a long period of time (i.e., draws a straight line at the top), one doesn't know if the oxygen is at 5.1 or at 15%. In short, an operator under these circumstances wouldn't know what the combustion conditions are. It is therefore good burning practice to always set the fuel rate and/or I.D. fan speed so that the oxygen recording is always at least below the 5% mark whenever fuel is being fired into the kiln. This recommendation also applies and is of equal importance during times of kiln starts and warm-ups. As a matter of fact, as soon as a main fire is lit in the kiln, it is the operator's first duty thereafter to stabilize the oxygen content at a level that is at least below the 5% oxygen mark.

16.3   FUEL-RATE CONTROL

Before any main fire (except the kiln warm-up torch) can be lit in the kiln, one has to ascertain that the following requirements are met:

a) Sufficient air is present to achieve complete combustion.
b) The gas analyzer is operational and functioning properly.
c) Sufficient heat is present to readily ignite the fuel.
d) The I.D. fan is running and the kiln draft is properly regulated to prevent an abrupt, delayed, explosive ignition of the flame.
e) The firing floor is cleared of any unauthorized personnel.
f) The primary-air fan is running and its flow rate properly set.

Once the fuel has ignited and a proper flame obtained, the operator must immediately check the gas-analyzer recording to ascertain that no combustibles (CO) are showing after a time delay of $\approx$ 30 s. After this, necessary adjustments (fuel or I.D. fan) to bring the oxygen (O_2) content below the 5% mark must be made. These are standard and elementary steps that apply to any kiln when lighting the main fire. Another rule that is practiced in most plants for safety reasons is to cut off the fuel completely whenever the flame is not lit after 30 s. The kiln is then purged for at least 5 min to rid the kiln of any combustible gases before another attempt is made to reignite the fire.

The operator should never attempt to control burning-zone temperature by merely increasing or decreasing the fuel rate, nor should he, for example, increase the I.D. fan speed to permit raising the fuel rate because the exit gas is deficient in oxygen. To operate in this manner is sure to lead to trouble, because no consideration is being given to the temperature at the back end of the kiln, where the feed is being prepared for calcination. Increasing the I.D. fan speed in order to be able to add more fuel (in case of low oxygen) will cause the back-end temperature to rise. This however is the wrong thing to do, because the back-end temperature plays a vital role in the operation of the kiln, and cannot be allowed to freely float up and down.

Whenever the operator makes a change in either the I.D. fan speed, the fuel rate, or both, he has to anticipate the possible reactions caused by the change, remembering that an increase in fuel rate results in higher back-end temperature, low percentage of oxygen in exit gas, and higher burning-zone temperature, and an increase in I.D. fan speed will result in higher back-end temperature, more oxygen in the exit gas, and lower burning-zone temperature. From this it becomes obvious that the fuel rate alone does not govern the burning-zone temperature. Without changing the fuel rate, a change in the burning-zone temperture can be caused merely by altering the kiln draft with I.D. fan speed. Nevertheless it should be emphasized that a fuel-rate adjustment will give the fastest reaction whenever a change in burning-zone temperature is required.

The first rule in fuel-rate control is: Always check the oxygen and CO-analyzer recording before and after any fuel-rate adjustment is made. It will not take long for a new kiln operator to learn that the burning zone will react slowly to any fuel-rate adjustment. There is an inherent time delay until a noticeable change in temperature takes place. Since these time delays can be as much as 10 min, particularly after an upset operating condition, it is quite common among new operators to either over-fuel or under-fuel the kiln for a prolonged length of time. The end result of this action is a burning zone that alternates from "cold" to excessively hot, a condition generally referred to as a cycling kiln. Because of this time delay, operators must become skillful in anticipating these swings of extreme temperatures and make their fuel adjustment before the actual change in temperature takes place. For proper timing of these adjustments the operator checks other instruments which could give him a positive sign that a turnaround in temperature conditions in the burning zone is imminent.

These so-called early warning signals are:

		Burning zone
Oxygen(%)—without fuel or fan adjustment	increases: decreases:	heating up cooling down
Kiln-drive torque	increases decreases	heating up cooling down
NO_x (%)	increases: decreases:	heating up cooling down
CO_2 (%)	increases: decreases:	cooling down heating up
Back-end temperature*	was higher than optimum: was lower than optimum:	heating up cooling down

* Refers to the temperature conditions that prevailed at a time that is equal to the travel time of the feed through the kiln, e.g., if travel time = 1.5 h, the operator looks at the back-end temperature that prevailed 1.5 h before.

The time lag between burning-zone temperature and a fuel-rate adjustment has been previously mentioned. With direct-fired coal systems this lag time is even more pronounced since it takes several minutes between the time the coal rate to the mill is adjusted and the change to be noticed at the burner tip.

Much work has been done with automatic fuel-rate control. On precalciner kilns, the fuel rate to the flash calciner is usually automatically controlled in a closed loop that uses the fourth-stage, preheater cyclone exit-gas temperature as the input variable. Simply stated, when this temperature is below the setpoint, more fuel is given to the flash calciner and, conversely less fuel when the temperature is above this setpoint.

Constraints in this loop are the same as with the fuel-rate adjustments at the rotary-kiln firing end, namely the oxygen must conform to certain predetermined levels and no CO is allowed to show. Since a flash furnace is much smaller than a rotary kiln, the controller must be tuned so that there will be a fast response by the coal feeder whenever a change in this fourth-stage, cyclone outlet temperature takes place.

Conventional fuel-rate control on wet, dry, and preheater kilns can be manually or automatically controlled. When automatic control prevails, the control concept should preferably consider as input variables at least the burning-zone temperature, the back-end temperature, and the percent oxygen in the kiln exit gases. A simple control logic of adjusting the fuel rate strictly and solely based on the burning-zone temperature alone seldom yields satisfactorily stable and efficient kiln operations.

During times of severe kiln upsets, i.e., when underburned fine clinker has entered the cooler and "blackened" out the burning zone, an operator often tends to make the mistake of overfueling the kiln. This is a natural tendency of all operators because a lot of heat is needed to bring the burning-zone temperature back to normal. Since the kiln speed is drastically reduced in such instances and visibility is extremely poor, they tend to leave the fuel rate at high levels for too long a period of time waiting for the burning zone to clear. However the kiln could already be in an over-heated condition. Things to consider during these conditions are:

a) Operating the kiln with no or very little oxygen contents (even when no combustibles are showing) often does not produce the desired heating of the burning zone. *Reason:* The dust in the kiln and low oxygen combined together produce lower flame temperatures. *Solution:* Try a little less fuel to operate at a slightly higher oxygen content of, e.g., 0.5–0.9%.

b) Once a heavy onrush of material has passed through the burning zone, there invariably is a lighter load behind it that causes a rapid increase in temperature when it arrives in the burning zone. This rapid temperature increase can usually not be detected early enough because of the dusty material in the cooler. *Solution:* Watch for a definite sign that the load is becoming lighter (oxygen increases, NO_x increases, kiln amp increases, cooler undergrate pressure decreases) and start to increase the kiln speed at that time. Depending on the severity of this light load it might become necessary to start reducing fuel rate at this time also (the effect on back-end temperature should not be forgotten) to prevent the burning zone from becoming too hot. In short, this is a time where anticipation of upcoming changes becomes of primary importance.

16.4 KILN-SPEED CONTROL

A kiln can never be expected to operate in a stable condition for an indefinite length of time because changes do take place in the kiln regardless of how stabilized the operation may appear. Sooner or later an adjustment in kiln speed has to be made if the kiln is to continue to produce a good clinker. The commonest action the operator must take in case of an apparent upset is the so-called "slow-down" procedure, as described in the following paragraphs, and shown graphically in Fig. 16.2.

After the kiln has been operating for a long period of time under balanced conditions the operator detects a heavy onrush of feed directly behind the flame. Whenever this happens, the operator must determine whether he will be able to hold the heavy feed load with an increase in the fuel rate alone, or are conditions such that the dark feed could pass too far under the flame even with the fuel rate increase? It is important that this decision be made at the earliest time possible. Oxygen percentage in the exit gas, back-end temperature, and conditions in the cooler are the deciding factors to be considered.

Assume that all indications are that it will not be possible to maintain the same kiln speed, and therefore speed will have to be reduced. It then becomes necessary to determine how much the kiln can be slowed. Depending on the magnitude of the "push," 5–10 rph less will be sufficient in some instances. In other cases the onrush is so heavy that the kiln has to be slowed down to minimum speed. Only experience will tell the operator how much the kiln has to be slowed. One rule, however, applies at all times: Never allow raw, unburned feed to enter the cooler, even if this means that the kiln has to be stopped and reheated on quarter turns or on auxiliary drive.

After the kiln has been slowed, there are, of course, a number of changes that take place, and the operator must be alert to watch for these changes and take appropriate corrective action. First, the back-end temperature will start to increase. This is undesirable because this temperature should be held within ± 20 degrees. Reducing the I.D. fan speed will aid in holding this temperature. Because of the decrease in the kiln draft resulting from the I.D. fan-speed reduction, there will be insufficient oxygen available for complete combustion. This is corrected by reducing the fuel rate until the exit-gas analyzer again indicates from 0.4–0.8% oxygen in the gas.

Because of the slower kiln speed, less material will enter the cooler, thus both secondary air temperature and undergrate pressure will decrease. Therefore, reduce the bedgrate speed in the cooler to maintain more or less the same undergrate pressure. This will be difficult to do. On very slow kiln speed it is almost impossible to obtain the same undergrate pressure as on full speed. Under slow-speed operation, do not attempt to hold the secondary air temperature on the same level as on full kiln speed, especially when the clinker is slightly underburned, because of the danger of the clinker not being sufficiently cooled when it leaves the cooler.

While the kiln is on slow speed it is necessary to decide how soon the speed can be raised again. First of all, never increase the kiln speed before there is a definite sign of the burning zone warming up, or there is a definite indication that the burning zone will warm up within a few minutes because of a lightening of the load. Do not hold the kiln on very low speed until the burning zone has reached normal clinkering temperature; start increasing the speed as soon as there is a sign of warming up. Signals and instruments to watch for these early warnings are discussed in Fig. 16.3. The more the kiln speed is increased, the smaller the speed increments should be and the longer the time interval between each speed increase. The way in which the kiln speed is increased is probably the most important factor in getting the kiln into stable conditions again. Consider the three examples shown in Fig. 16.3.

Example 1: This shows an ideal execution of a speed-up procedure. The operator started to increase the kiln speed in large steps of 5 rph at a time until he reached kiln speed 50. From then on he extended the time between each speed increase and carried out the increases in smaller steps as the speed was increased. These were at the rate of 3 rph from 50 to 62, 2 rph from 62 to 66, and 1 rph from 66 to full speed of 70 rph.

Example 2: In this example the operator used the wrong judgment. He started to raise the kiln speed too fast before the burning zone was ready for it. Very soon he had to backup on the kiln speed and had to start reheating the burning zone all over again.

Example 3: Here the opposite condition occurred. The operator waited too long on slow kiln speed. Suddenly the burning zone started to gain heat very rapidly, forcing him to raise the kiln speed in large incre-

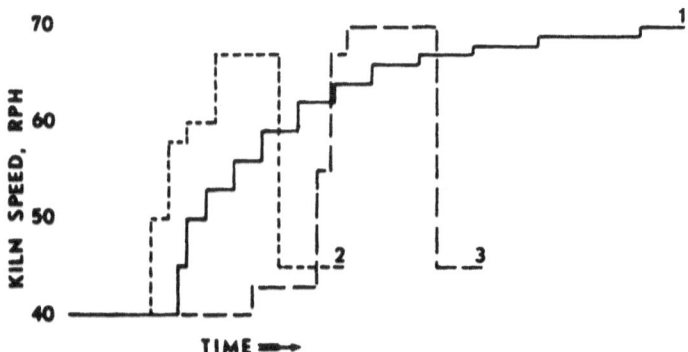

Fig. 16.3 Kiln speedup, for any reason, must be done in easy stages, as shown in Example 1, commencing with large steps close together, then extending the time between steps as the steps become smaller.

ments and in short time intervals. For a short time full kiln speed was able to be maintained, but very soon the kiln became overloaded and the operator had to slow the kiln down once again.

Remember that every time the kiln speed is increased, the back-end temperature drops, requiring more fuel to burn with each speed increase. This requires an adjustment in the I.D. fan speed as well as in the fuel rate. Both are increased in a manner similar to the kiln-speed increase in Example 1 (Fig. 16.3).

The ideal example given in Example 1 by no means suggests that the kiln will operate in a stable manner again for a long time on full speed. In time intervals of 2–3 h another push, each time less severe, will be encountered. Each time the slow-down period will be shorter and each time the chances are better that the kiln can be held at the normal operating speed. Sometimes one slow-down will be sufficient, at other times two or three such slow-down sequences have to be undertaken until the kiln again is stable. The chances of avoiding a cycle in the kiln are minimized when the operator executes a slow-down sequence according to the ideal example given. The aforementioned discussion applies primarily to long-wet and dry kilns. Preheater and precalciner kilns react differently.

Kiln Rollback. The feed bed in a rotary kiln occupies up to 10% of the cross-sectional area of the kiln. While the kiln is rotating, the center of gravity of this material is displaced to one side (on the rising side) of the 6 o'clock position (bottom center of the cross section), the torque resulting from this displacement being opposed by the driving torque applied to the kiln. When power to the·drive motors is cut off for any reason, the off-center position of the load causes the kiln to rotate in the reverse direction until the feed bed comes to rest at the 6 o'clock position. The reverse rotation of rollback can reach a maximum speed that exceeds the normal running speed of the kiln, thereby damaging the drive gear.

Another problem associated with rollback is that of restarting a kiln that has stopped while loaded. If left to rotate freely after a stop, the kiln will come to rest with the load of the feed bed at or near the 6 o'clock position. In the past the relatively small rotary kilns could be started from this position without much difficulty, as the kiln drives then in use were capable of overcoming the initial torque to set the kiln in motion. With the advent of the 500-foot or longer giants, kiln manufacturers had to find a way to overcome static fricton and inertia in order to accomplish initial acceleration with a minimum of strain on the drive gear and motor. This resulted in the concept of kiln rollback control. In this method of control, instead of permitting the kiln to reverse rotation, or roll back, when the drive motor is stopped, a brake is applied to the drive unit to stop the kiln at the instant the motor circuit is broken with the feed-bed position in an inclined angle to the rising side. Thus the gravitational force of the kiln feed bed acting downard is frozen by the brake so that it can be used advantageously when the kiln is started again.

When the operator presses the kiln-drive start button, the following sequence of actions takes place automatically; first the brakes are released permitting the kiln to roll backwards (Fig. 16.4) with the feed bed past the low point of the kiln circle. Next the feed bed, now being on the down side of the kiln, causes the kiln to start to roll, or rotate, forward in the normal direction of rotation, at which time the kiln drive motor is activated to keep the kiln revolving in the forward direction.

This system has proved its usefulness and has contributed significantly to the present long life of drive gears and motors of large rotary kilns.

The most commonly asked question is: "What is the optimum kiln speed for efficient and stable kiln operation?" There is no clear-cut answer because each kiln is different in its requirement. The kiln slope and the resultant feed velocity in the kiln is one factor that must be considered. Another important factor is the feed loading of the kiln, i.e., the output rate that can be achieved at a given kiln speed without sacrificing operating stability. Typically, specific volume loading of kilns varies from a low of

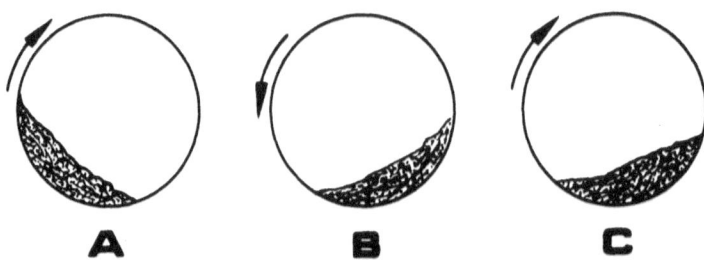

Fig. 16.4 Kiln rollback is used to advantage when starting a kiln under load. When the drive motor is stopped, brakes are applied to hold the feed bed in Position A. The arrow indicates the normal direction of rotation. When the drive-motor start button is pressed to start a loaded kiln, the brake is automatically released, permitting the kiln to roll back to Position B. The kiln now starts to roll forward as shown in Position C, at which time the motor is activated to continue rotating the kiln.

approximately 5% to a high of 10%.

Dry and wet kilns with steep slopes should be operated at slower speeds to prevent large onrushes of feed waves into the burning zone (and into the cooler) during upset conditions. Such kilns usually operate normally in the range of 50–60 rph. Other less-sloped dry- and wet-process kilns generally run at speeds of 60–72 rph whereas preheater and precalciner kilns operate at much higher speeds. Kilns that consistently and periodically exhibit difficult burning conditions in the burning zone as a result of incomplete feed calcination sometimes exhibit better operating stability when the kiln-speed target is lowered and the kiln is concurrently operated at a higher percent specific loading. This, however, is in direct conflict with the "old" heat-transfer law for rotary kilns which states that more efficient and favorable heat-transfer conditions exist when the kiln is operated at high kiln speed and low percent feed loading. Hence, instead of lowering the kiln-speed target and operating at a deeper bed depth to overcome frequent operating instability, it might make more sense to equip such kilns with more heat exchangers in the calcining zone to improve the overall heat transfer behind the burning zone. Regardless of the kiln-speed target (top speed under normal operation) it is good practice to set this target at a speed that is approximately 5–15 rph lower than the maximum permissible. The kiln must be allowed some "elbow room" at the top to make it possible to further increase the kiln speed in an emergency when the burning zone has become dangerously overheated.

17.

Fuel Systems

17.1 FUEL HANDLING AND COAL GRINDING

Cement kilns are usually fired with oil, natural gas, or coal. Gas firing requires no fuel preparation; this type of fuel is used directly as it is being delivered to the plant by the gas pipe line. In oil firing, the oil has to be preheated to a given temperature that produces the desired viscosity for proper atomization of the fuel at the burner tip. This is usually accomplished in heat exchangers that predominantly employ steam as the heating medium. Automatic controls must maintain the oil temperature within a very narrow range of ± 5 C (± 9 F) which is considered essential for uniform firing conditions and flame stability.

With *gas firing*, the operator's duty is relatively simple in matters of fuel-flow monitoring. The operator must watch out for the possibility of sharp pressure changes in the gas supply line that would indicate some problems along the supply line. Gas firing of the flash furnace in a precalciner kiln, however, has been shown to present some problems when compared to the other two fuels. Here, combustion conditions have to be very closely monitored to ensure that all the gas introduced into this auxiliary firing unit is completely burned within its confines. When this prerequisite is not fulfilled, there exists the danger that some of the unburned natural gas could escape into the upper stages of the preheater cyclones, undergoing combustion there and ultimately causing elevated temperatures of the gases leaving the preheater tower.

With *oil firing* attention must be paid to the oil preheat temperature as any significant change could lead to a change in flame characteristics. Furthermore, close attention must be given to the pressure indicators in the heat exchangers because plugged filters could lead to an interruption of

253

the oil flow to the burner. Steam pressure must also be frequently monitored to make sure sufficient steam is available to preheat the oil properly.

Control of the *coal-grinding plant* and coal conveyance to the burner is much more demanding than when either one of the above-mentioned liquid or gaseous fuels is used. An operator, for safety reasons, should not be allowed to exercise control over the coal handling and grinding plant unless that operator is completely familiar with all safe operating procedures for the system. Coal firing has its advantages, namely, a coal fire is more luminous and delivers better heat transfer by radiation from the flame to the feed. In short, there is less time lag between the moment a fuel adjustment has been made and the time the clinker bed reacts to this change. However, this advantage is neutralized in direct-fired kilns because of the time delay between the moment coal-rate adjustments are made and the coal has been ground and insufflated into the kiln. Operators that have had the opportunity to fire different kinds of fuel tend to agree that the burning-zone temperature can be more rapidly adjusted with a coal fire than with any other type of fuel.

An entire book could be written on the subject of coal handling, grinding, and firing. For the benefit of the operator, the subjects that are of paramount importance to him, namely his safety and the safety of the equipment, are discussed here. An operator can become comfortable with coal firing provided that a respect for its limitations and inherent safety requirements is developed. The operator must know the system's idiosyncrasies and immediately be able to recognize a potentially dangerous condition when it develops. Pulverized coal, mixed with excessive air and exposed to excessive temperatures, represents a potentially explosive mixture. High-volatile coal, even in the unground state, can also undergo spontaneous ignition while in storage. Prerequisites for successful and safe grinding and handling operations for coal are:

a) well-established, safe standard-operating procedures,
b) close and continuous attention to the instrumentation by competent operators, and
c) regular preventative maintenance for all system components.

Even the best-designed and maintained systems will sooner or later present the operator with an unusual condition that demands immediate and corrective action on his behalf. An operator must be on the lookout for the following *potentially dangerous conditions* each of which could lead to a fire or explosion:

- Fresh smoldering coal (i.e., coal that has started spontaneous ignition) being fed to the coal mill
- Metal pieces, rags, and other materials in the coal being fed to the mill causing mill, bin, or feeder plug-ups and metal fragments not being ejected from the mill (sparks could ignite the coal)
- Mill outlet temperature too high or showing a rapid rise
- Fresh unground coal too wet (> 15% moisture) that could lead to any one of the following conditions:
    a)   insufficient drying in the mill,
    b)   accumulations in the mill, bins, and/or coal pipe,
    c)   excessively high mill inlet-temperature demands, and
    d)   too low a mill outlet temperature
- Insufficient air velocity in the mill and coal pipes causing settlement and accumulations of coal or possible entrance (back flashing) of hot kiln gases into the burner pipe
- On semidirect or indirect-fired kilns: worn vanes on rotary feeder or malfunctioning air locks—allowing either coal to seep into the primary air pipe or hot air to enter the coal bin during shutdowns
- Leaks in the coal pipe
- Poor housecleaning with large accumulations of coal near the coal handling, storage, or grinding system and on the firing floor
- Entire coal system or primary air fan has shut down under load due to a power failure, i.e., coal is present in the mill and the burner pipe (IMPORTANT; DO NOT OPEN DOORS OF SYSTEM FOR MANUAL CLEANOUT OF ACCUMULATIONS BEFORE THE SYSTEM HAS COOLED DOWN TO AMBIENT TEMPERATURE)
- Ground coal insufficiently dried causing settlement, plugging, and accumulations (coal pockets) in bins, dust collectors, and coal pipes
- Entire coal system has not been properly cleaned (air swept) when the kiln was stopped for a prolonged shutdown
- Operating the system with faulty instrumentation, improper damper control, and lack of proper fire-extinguishing and explosion-prevention equipment or devices
- Smoking by employees or welding by maintenance crew near the system while it is in operation

An operator should discuss all of these conditions with his supervisor and must know what corrective action should be taken if any of these conditions arise. These procedures must be clearly and fully memorized because there will not be sufficient time available to look these up in the standard operating procedures when such emergencies develop.

When coal is received at the plant, it has to be dried as most coals contain appreciable amounts of moisture. This drying is usually done in the coal mill itself using hot excess cooler air. This hot cooler air is tempered with cool ambient air a short distance past the cooler to prevent:

a) overheating of the dust cyclone and air pipe ahead of the coal mill, and
b) excessively high temperatures from entering the coal mill when overheated conditions prevail in the cooler itself.

Other plants use the inert preheater exit gases for drying in the coal mill. Since these gases usually contain less than 5% oxygen, they represent a safety advantage when compared to the use of cooler excess air.

IMPORTANT:

a) Know what the maximum allowable temperature is for the coal mill *inlet* temperature.
b) Do not exceed this temperature under any conditions.
c) Make sure that this tempering damper is not fully open during normal operating conditions to allow for further admittance of cold air when the need arises.

A second tempering damper is usually installed just ahead of the coal mill that is the primary control for the coal mill *outlet* temperature. Cold ambient air is drawn into the system to hold this coal mill outlet temperature within a narrow range regardless of changes in the moisture content of the unground coal. Typical setpoints for coal mill outlet temperatures are:

for low-volatile coal:	80–90 C (175–195 F)
for high-volatile coal:	68–80 D (155–175 F)

When this coal mill outlet temperature exceeds 93 C (200 F) the danger of premature ignition of the coal in the mill exists.

Grinding of the coal is done predominantly in roller mills although there are still several plants that use ball mills for this purpose. Both these types of mills rely on air-sweeping action to evacuate the coal from the mill. In semidirect- and indirect-fired systems the coal/air mixture is blown into a holding bin whereas in direct-fired systems the coal is directly insufflated into the burning zone. **IMPORTANT:** The following fundamental conditions must always prevail whenever any of these systems are in operation:

a) Sufficient air has to pass through the mill to properly evacuate the ground coal from the mill. This can be monitored by the draft at the mill inlet and the mill outlet. Both these drafts must not be allowed to drop below a given preestablished value. Low-pressure differentials between these two measuring points are an indication of mill plug up or lack of sufficient air for sweeping the coal from the mill. An operator should know exactly what the critical values for coal mill drafts are for the particular system under his control. The same principle applies to dust collectors incorporated into an indirect-firing system.

b) To keep the coal particles in suspension while being transported from the mill to the bin or burner, a minimum velocity of air passing through the coal pipes must exist at all times. As a rule of thumb it is advisable *not* to let this velocity drop below 30 m/s (6000 ft/min). Since there is a direct relationship between the mill-exit draft and this velocity, an operator monitors this draft (on direct-fired kilns this is the primary air pressure) and must know what the minimum permissible pressure is. The operator will then have to make sure to never operate below this critical pressure whenever coal is being ground in the mill.

c) Since one kilogram (or pound) coal requires approximately 8.5–10.5 kg (or lb) air for combustion, the fuel/air mixture emanating from the coal mill must be kept well below this critical ratio to prevent premature ignition. Most coal mills operate at air/fuel ratios that are less than 2.0 kg (lb) per kg (lb) of coal. However, direct-fired kilns, with their high percent of primary-air requirements, usually operate at higher ratios and therefore need closer attention.

d) Combustion air (primary, secondary, or tertiary air) should be
   high enough to ensure continuous ignition of the fuel or light-up
   torches must be lit until the fuel can maintain self-ignition.

All coal-handling and grinding facilities should be equipped with
automatic shut-down interlocks. Some of the more common interlocking
systems are:

Operating Condition	Automatic Action Taken
1. I.D. fan stopped	All fuel off
2. Combustibles for longer than 1 min	All fuel off and precipitator deenergized
3. Gas temperature too high in precalciner	Flash furnace shutdown
4. Primary air fan stopped	Fuel off
5. Temperature too high at mill outlet	Coal mill shutdown and activation of CO_2 extinguishers
6. Flash furnace temperature too low	Fuel in flash furnace off
7. Cooler exhaust fan stopped	Fuel off
8. Pulverized coal bin shows rapid temperature rise	Activate inert gas (CO_2, N_2) and fire suppressants

The optimum coal fineness must be established for each kiln and depends
on such factors as type of coal fired and type of flame required. Subbituminous and high-volatile bituminous coal are usually ground coarser
than low-volatile coal or coke. As a general guideline, the following
criteria can be used whenever the coal characteristics change:

If	Then
Volatilles increase (%)	Grind coarser
Volatiles decrease (%)	Grind finer
Ash increases (%)	Grind finer
Ash decreases (%)	Grind coarser
Ash rings form at kiln outlet	Grind finer (avoid oversized particles)
Plume too long (late ignition)	Grind finer and/or increase secondary air temperature
Plume too short (early ignition)	Grind coarser and/or decrease primary air temperature

Another factor to consider is the so-called ash-softening temperature as this could be a primary cause for troublesome ring formations at the kiln discharge. It is generally believed that coals with a low ash-softening temperature and/or high contents of iron in the ash are prime contributors to such ring formations. Here, too, it is recommended that finer grinding be employed to combat such rings.

17.2 FUEL BURNERS AND FLAMES

It is difficult to control the shape of a coal flame during the course of an operation unless one of the modern, sophisticated adjustable burners is used. A change in primary air pressure is about the only real adjustment an operator can make while the kiln is in operation. Although position and the design of the burner as well as hood draft and secondary air temperature are prime influencing factors for flame shape, these variables can not be easily changed by the operator.

Very complex adjustable and/or combination burners are successfully used for oil, gas, or coal firing of cement kilns. These are excellent tools for flame and coating control but are often misused. Many times operators have opposing views about appropriate and desirable flame shapes and thus make too many adjustments too frequently. This is especially true with burner designs that are equipped with adjustable inserts to promote turbulence and mixing of air with the fuel. Furthermore, any coal burner that contains such inserts lends itself to potential coal-pocket formation during unusual operating conditions. Because these adjustable burners are so efficient in flame-adjustment capabilities, any error in adjustment setting by the operator can directly lead to thermal abuse of the refractory and kiln components near the kiln discharge area. This has compelled many plants to use a straight burner pipe with no inserts for natural gas or coal firing. However, the advantages in low costs and prevention of coal-pocket formation is outweighed by the disadvantages and limitations of these simplified burners when used on modern preheater and precalciner kilns. Straight burner pipes are designed for normal "full-speed" production operation and usually do not deliver satisfactory flame shapes and characteristics during start-up and kiln slow-down periods. During these times the flame is often too long and the ignition of the fuel erratic causing dislocations in the position of the burning zone. Long, lazy flames can also create excessive coating and ring formations. Operators are usually capable of coping with this problem by using the so-called auxiliary start-up torch as an aid in ignition which by itself is expensive and requires frequent watching. Short-term dislocations of burning-zone positions in wet- and dry-process kilns are not so critical but can become disastrous on preheater and precalciner kilns. On these modern kilns the burning-zone position can not be allowed to fluctuate as freely because these kilns operate at much higher speeds and are considerably shorter in length. Another disadvantage of the straight burner pipe is the need for redesigning the burner geometry (for proper tip velocity) whenever there is a major modification made on the kiln that would produce a significant change in the specific fuel requirements.

All fuel burners are exposed to severe wear and heat conditions inside the kiln and should therefore be frequently inspected for thermal damage by the maintenance and operating personnel.

17.3 TESTING COAL BURNERS FOR TIP VELOCITY

The following is a method to determine the actual tip velocity on coal burners when no actual flow measurements are possible.

Data Needed

		Metric	English
W = optimum kiln output, steady state		___kg/h	___sh.t/h
H = average specific fuel consumption		___kJ/kg	___Btu/sh.t
A = fuel heat value (as fired)		___kJ/kg	___Btu/lb
M = percent moisture in coal (mill inlet)		___(decimal)	
V = percent volatiles in coal (as fired)		___(decimal)	
T = primary air temperature		___C	___F
p = burner-tip pressure		___mm H_2O	___in. H_2O

Calculations:

1. *Weight of dry fuel fired per minute:*

$$w_1 = \left(\frac{W}{60}\right)\left(\frac{H}{A}\right) \qquad = \underline{\qquad} \text{kg/min (lb/min)}$$

2. *Total combustion air required (@ 5% excess air):*
   a) English units

$$Q_1 = \left(\frac{A}{12600}\right) 10.478W \qquad\qquad = \underline{\quad\quad} \text{lb air/min}$$

   b) Metric units

$$Q_1 = \left(\frac{A}{29290}\right) 10.478W \qquad\qquad = \underline{\quad\quad} \text{kg air/min}$$

3. *Primary air flow:*

$$a = (x)(Q_1) \qquad\qquad = \underline{\quad\quad} \text{kg/min (lb/min)}$$

   where:
   x = % primary air
   Note: In selecting the proper primary air flow, one has to consider the amount of air needed to evacuate the coal from the mill when the direct-firing method is used. Also important is the fan static-pressure rating.
   Guidelines: x = 0.16 – 0.23 for direct-fired kilns
   x = 0.07 – 0.14 for semidirect- and indirect-fired kilns

4. *Volume of primary air:*
   a) Determine density (d) of primary air
      English units

$$d = 0.080714 \left(\frac{492}{T + 460}\right)\left(\frac{14.7 + 0.0361p}{14.7}\right) = \underline{\quad\quad} \text{lb/ft}^3$$

      Metric units

$$d = 1.2929 \left(\frac{273.2}{T + 273.2}\right)\left(\frac{760 - 0.0736p}{760}\right) = \underline{\quad\quad} \text{kg/m}^3$$

   b) Determine volume (v_1) of dry primary air

$$v_2 = \frac{a}{d}$$

$$v_1 = \frac{a}{d} \qquad\qquad = \underline{\quad\quad} \text{m}^3/\text{min (ACFM) dry}$$

   c) Determine the water vapor (v_2) from moist coal
      (direct-fired systems only)

$$v_2 = wMZ \qquad\qquad = \underline{\hspace{2cm}} m^3/min \text{ (ACFM) vapor}$$

where:

Z = density of water vapor (m^3/kg, ft^3/lb)

Guidelines for Z:							
	English units				Metric units		
for T (°C)	71	82	93	T (°F)	160	180	200
Z =	4.81	3.12	2.06	Z =	77	50	33

Note: The selection of the proper mill-outlet temperature should take into consideration the volatile content of the coal. As a rule of thumb use:

If V = > 35%,   then recommended T = 65 C (150 F)
          = 25–35%   = 71 C (160 F)
          = 12–25%   = 82 C (180 F)
          = 5–12%   = 90 C (195 F)
          = < 5% (petroleum coke)   = 102 C (215 F)

d) Determine the actual volume of moist primary air

$$Q_2 = v_{-1} + v_2 \qquad\qquad = \underline{\hspace{2cm}} m^3/min \text{ (ACFM) moist}$$

5. *Tip velocity:*

Area = 3.1416r

where:

r = radius of burner tip

$$\text{Velocity} = \frac{Q_2}{\text{Area}} \qquad\qquad =- \underline{\hspace{2cm}} m/min \text{ (ft/min)}$$

$$\frac{0.016667Q_2}{\text{Area}} \qquad\qquad = \underline{\hspace{2cm}} m/s \text{ (ft/s)}$$

Using the above method in computing tip velocity, the characteristics of coal burners have been determined and are shown in Table 17.1 (direct-fired, wet-process kiln), Table 17.2 (direct-fired, dry-process kiln), and Table 17.3 (indirect-fired, suspension preheater kiln).

TABLE 17.1

BURNER-TIP VELOCITY ON COAL BURNERS
(Direct-Fired Wet-Process Kiln)

Data Input		English		Metric	
W	= optimum kiln output, steady state	32	sh.t/h	29030.4	kg/h
H	= average specific heat consumption	505000	Btu/sh.t	5873.655	kJ/kg
A	= fuel heating value (as fired)	12300	Btu/lb	28622.1	kJ/kg
M	= percent moisture in coal (mill inlet)	0.11	(decimal)	0.11	(decimal)
V	= percent volatiles in coal (as fired)	0.28	(decimal)	0.28	(decimal)
T	= primary air temperature	180	F	82	C
p	= burner-tip velocity	5.1	in. H_2O	1295	mm H_2O
x	= percent primary air	0.21	(decimal)	0.21	(decimal)
c	= diameter of burner tip	0.96	ft	10.292	m
	effective burner-tip opening	0.7213	ft²	0.0670	m²
1.	Weight of dry fuel fired per minute	219.0	lb/min	99.3	kg/min
2.	Total combustion air required	2239.7	lb/min	1016.6	kg/min
3.	Primary air flow	470.3	lb/min	213.5	kg/min
4.	Volume of primary air a) density	0.0628	lb/ft³	0.8691	kg/m³
	b) volume air	7486.5	ACFM dry	245.6	m³/min
	c) water vapor	1204.3	ACFM	34.1	m³/min
	Total primary air	8690.8	ACFM	279.7	m³/min
5.	Tip velocity	12049	ft/min	4174	m/min
		201	ft/s	70	m/s

TABLE 17.2

BURNER-TIP VELOCITY ON COAL BURNERS
(Direct-Fired Dry-Process Kiln)

Data Input		English		Metric	
W =	optimum kiln output, steady state	73	sh.t/h	66225.6	kg/h
H =	average specific heat consumption	4100000	Btu/sh.t	4768.71	kJ/kg
A =	fuel heating value (as fired)	11980	Btu/lb	27877.46	kJ/kg
M =	percent moisture in coal (mill inlet)	0.13	(decimal)	0.13	(decimal)
V =	percent volatiles in coal (as fired)	0.32	(decimal)	0.32	(decimal)
T =	primary air temperature	180	F	82	C
P =	burner-tip velocity	5.6	in. H_2O	1422	mm H_2O
x =	percent primary air	0.18	(decimal)	0.18	(decimal)
c =	diameter of burner tip	1.2500	ft	0.3810	m
	effective burner-tip opening	1.2272	ft²	0.1140	m²
1.	Weight of dry fuel fired per minute	416.4	lb/min	188.8	kg/min
2.	Total combustion air required	4148.2	lb/min	1882.9	kg/min
3.	Primary air flow	746.7	lb/min	338.9	kg/min
4.	Volume of primary air　a) density	0.0629	lb/ft³	0.8569	kg/m³
	b) volume air	11870.5	ACFM dry	395.5	m³/min
	c) water vapor	2706.5	ACFM	76.6	m³/min
	Total primary air	14577.0	ACFM	472.1	m³/min
5.	Tip velocity	11878	ft/min	4141	m/min
		198	ft/s	69	m/s

TABLE 17.3

BURNER-TIP VELOCITY ON COAL BURNERS
(Indirect-Fired Suspension Preheater Kiln)

Data Input		English		Metric	
W	= optimum kiln output, steady state	89	sh.t/h	80740.8	kg/h
H	= average specific heat consumption	2950000	Btu/sh.t	3431.145	kJ/kg
A	= fuel heating value (as fired)	11980	Btu/lb	27877.46	kJ/kg
M	= percent moisture in coal (mill inlet)	0	(decimal)	0	(decimal)
V	= percent volatiles in coal (as fired)	0.26	(decimal)	0.26	(decimal)
T	= primary air temperature	180	F	82	C
P	= burner-tip velocity	8.5	in. H_2O	2159	mm H_2O
x	= percent primary air	0.09	(decimal)	0.09	(decimal)
c	= diameter of burner tip	0.67	ft	0.203	m
	effective burner-tip opening	0.3491	ft²	0.0324	m²
1.	Weight of dry fuel fired per minute	365.3	lb/min	165.6	kg/min
2.	Total combustion air required	3638.9	lb/min	1651.7	kg/min
3.	Primary air flow	327.5	lb/min	148.7	kg/min
4.	Volume of primary air a) density	0.0633	lb/ft³	0.7860	kg/m³
	b) volume air	5170.2	ACFM dry	189.1	m³/min
	c) water vapor	0.0	ACFM	0.0	m³/min
	Total primary air	5170.2	ACFM	189.1	m³/min
5.	Tip velocity	14811	ft/min	5832	m/min
		247	ft/s	97	m/s

18.

Clinker Cooler Control

It is extremely important for the operator to master cooler controls as well as other control functions of the rotary kiln system, because the cooler itself performs an integral part of the clinker-burning process. Cooler conditions influence the burning process in the kiln and consequently, the quality of the clinker. There are many instances where cooler operating problems can lead a perfectly stable kiln into a severely upset condition with possible damage to the equipment.

With the introduction of large-sized rotary kilns having production capacities in excess of 1500 tons per day, traveling grate coolers have become a problem in some plants. Unexpectedly, widespread complaints began as cooler components burned up and entire coolers had to be repaired at alarming rates. These problems were not only experienced in North America but were also encountered throughout the world wherever large rotary kilns had been placed into service. Regrettably, a final solution to the problem has not yet been found, and in some cement plants frequent cooler failures due to overheating still persist. On the other hand, major modifications and improvements in the design of cooler equipment in the past few years have lessened the problems to a large extent and indications are that coolers can now be made to operate in a satisfactory manner. Also, not all cooler failures are caused by design of the cooler itself, for inexperience and laxness on the part of operators has contributed to the problem. The detailed description of cooler control functions in this chapter can give the operator the information he needs to help lessen the frequency of cooler failures.

Fig. 18.1 shows a schematic layout of cooler controls to familiarize the

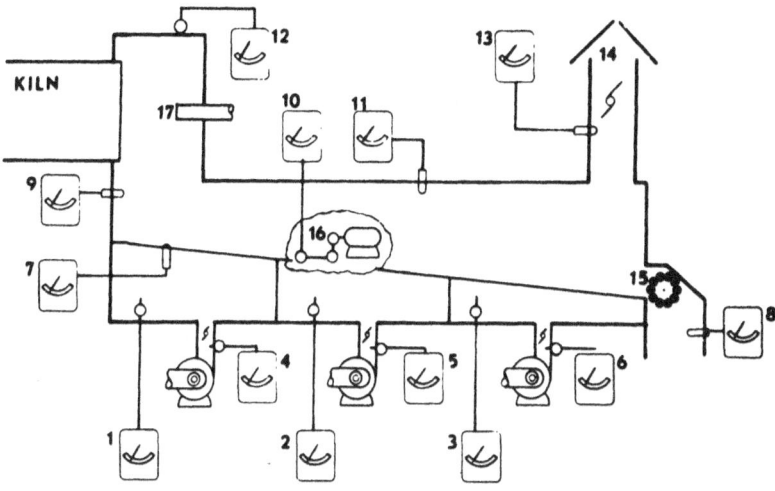

Fig. 18.1 Sectional diagram of a traveling-grate cooler. Instrumentation is as follows: 1, undergrate pressure, 1st compartment; 2, undergrate pressure, 2nd compartment; 3, undergrate pressure, 3rd compartment; 4, Fan No. 1 differential pressure; 5, Fan No. 2 differential pressure; 6, Fan No. 3 differential pressure; 7, cooler grate temperature; 8, clinker temperature at cooler outlet; 9, secondary air temperature; 10, grate drive speed; 11, cooler air temperature; 12, hood pressure; 13, cooler outlet gas temperature; 14, excess air chimney; 15, clinker crusher; 16, drive unit; 17, burner.

reader with the various terms used. It does not represent any particular make of cooler, but shows the components and principles involved in cooler construction and operation.

18.1 CRITICAL VARIABLES IN COOLER CONTROL

The kiln operator must operate the cooler in such a manner as to meet the following objectives as closely as possible:

a) Clinker temperature at the discharge of the cooler should be as low as possible because high temperatures endanger the transport equipment and waste valuable heat.

b) Secondary air temperature should be as stable and high as possible because this is a prerequisite for overall kiln operating stability and good fuel efficiency.

c) Cooler exit-air temperature should be as low as possible and volume as small as possible to assure a minimal amount of heat wasted to the atmosphere.

d) Hood pressure should always be slightly negative.

e) Depth of the clinker bed in the cooler should be such that a free passage of air through the bed can take place.

f) Cooler control settings should be such that bedgrates, cooler drive unit, clinker crusher, and cooler walls cannot become overheated.

The kiln operator has basically two control variables for accomplishing the above-mentioned objectives: the speed of the bedgrates which alters the clinker residence time and the clinker bed depth in the cooler, or the air distribution in the cooler can be changed.

In case of an emergency, such as badly overheated cooler conditions, the kiln operator has two more possibilities to bring the cooler under control: the amount of clinker falling into the cooler can be decreased by slowing the kiln speed, or the temperature of the clinker falling into the cooler can be lowered by adjusting the flame geometry to shift the burning zone further back in the kiln.

Numerous indicators and recorders on the plant control panel provide the means by which the operator maintains surveillance over operation of the cooler and enables the detection of irregularities in operation. These instruments should be observed and checked on a regular basis. Those that are most commonly used are undergrate air pressure, secondary air temperature, grate speed amperage drawn by grate drive motor, and the clinker-discharge temperature at the cooler outlet. In addition to these five essential instruments, there are other recorders that assist the operator in obtaining an overall indication of cooler performance. These additional instruments record the grate temperature, cooler exit-air temperature or circulating air temperature, flow rate of air forced into the cooler, and quench-air temperature measured at a point midway of the cooler length above the clinker bed. Other frequently used instruments are a nuclear gauge for measuring clinker bed depth, and a television camera and monitor showing the cooler interior.

The discussion that follows focuses on traveling grate coolers since these are the most commonly used and the most difficult to control. This type of a clinker cooler is shown in Fig. 18.1.

18.2 UNDERGRATE PRESSURE AND AIR-FLOW RATE CONTROL

These two controls, undergrate pressure and air flow, are probably the most significant parts of cooler control because they constitute the key to successful achievement of the objective of cooler control. A thorough knowledge and understanding of these controls is essential to the operator to enable him to maintain operating stability of the kiln and prevent the cooler components from overheating.

Before the procedures in undergrate pressure control are explained, a very important point has to be stressed: The described procedures are only vaild when the cooler contains clinker that has been properly burned. These procedures should not be followed when unburned, dusty clinker or raw feed has entered the cooler.

The cooler system shown in Fig. 18.1 has three undergrate compartments, each compartment receiving cooling air from an individual fan. The cooler bed-grate drive unit is in the center portion of the cooler. For control purposes, various instruments record the undergrate pressure in each compartment, air-flow rates delivered by each fan, and the speed of the bed grates. Under normal operating conditons (stable operation), there is an undergrate pressure for each compartment that ensures proper cooling of the clinker. By holding this undergrate pressure constant, the operator will theoretically hold the cooler control fairly constant and in balance.

Undergrate pressure is governed mainly by the following factors:

a) Depth of the clinker bed over the grates
b) Average particle size of the clinker in the cooler
c) Temperature of the clinker in the cooler, and
d) Amount of air introduced into the cooler.

Thick beds have higher resistance and thus require more force from the fan to push the air through than a thin bed. In other words, assuming that the other factors listed above remain constant, a thicker clinker bed results in higher undergrate pressure.

Depth of the clinker bed can be controlled by speed of the cooler grates. When a thinner bed is required, the grate speed is increased. A deeper bed can be obtained by slowing the grate speed. Because of the relationship between undergrate pressure and bed depth, it is possible to maintain a constant undergrate pressure by regulating the grate speed. Some cooler installations have automatic controls that work on this principle; a setpoint is selected on the controller for a desired undergrate pressure and the grate speed is automatically adjusted whenever the pressure deviates from the setpoint.

This approach in cooler control is usually satisfactory when the kiln operates in a stable fashion. However, during upset kiln conditions, or when the clinker characteristics change, this approach leaves much to be desired. During such times, an operator can often experience difficulties in maintaining proper cooling of the clinker. For example, consider the base operating condition shown in Table 18.1 with a second compartment bed depth of 14 in. (36 cm) at an hourly clinker throughput rate of 73 tons. In this base case, the clinker residence time in the cooler is 20.9 min and the clinker-discharge temperature 224 F. Now assume that the kiln suddenly starts to discharge clinker at a rate of 89 tons/h and that the grate speed is on automatic control, i.e., starts to increase the speed to maintain the same clinker-bed depth (see Table 18.2). Increasing this grate speed in an attempt to maintain a constant undergrate pressure, will then result in a reduction of the clinker residence time in the cooler, as shown in Table 18.2, to 17.2 min. This example shows that by pushing the hot clinker at a faster rate through the cooler and maintaining the same specific air-flow rate and undergrate pressure, the clinker-discharge temperature will increase to a theoretical 317 F. The lesson to be learned from these examples is that whenever a significantly larger amount of clinker falls into the cooler and as a consequence the grate drive speed increases significantly, the operator must compensate for this with a corresponding increase in the specific air-flow rate (SCFM).

Other influencing factors that cause the undergrate pressure to change are the temperature of the clinker in the cooler and the amount of air introduced into the individual undergrate compartments. Assuming other factors to be constant, a higher temperature of the clinker results in a higher undergrate pressure and increased air-flow rates in the undergrate compartments yield higher undergrate pressures.

TABLE 18.1

COOLER OPERATING STUDY (BASE CASE)

			Compartment				
		1	2	3	4	5	Total
Length (ft)		1.5	5	12.5	20	24	63
Width (ft)		7	7	9	9	9	
Depth (ft)		1.50	1.17	0.91	0.91	0.91	
T.P.H. output		73	73	73	73	73	
Weight of clinker		1544	4002	10033	16052	19263	50893
Clinker residence time		0.6	1.6	4.1	6.6	7.9	20.9
Clinker input	°F	2500	2057	1363	640	353	
Air input	°F	90	90	90	90	90	
Air output	°F	2090	1520	840	420	240	
Lb air (residence time)	°F	350	1980	9800	14000	16500	42630
Clinker output	°F	2057	1363	640	353	224	
Air needed (SCFM)		7339	16013	31613	28225	27305	110494

MBtu/t heat consumption   4.1
Air needed in kiln (lb/min)   4077.76   = 54,234.3 SCFM
Excess cooler air (SCFM):   56260.1   = 50.9% of total air in

TABLE 18.2

COOLER OPERATING STUDY (22% more clinker in cooler but air-flow rate the same)

			Compartment			
	1	2	3	4	5	Total
Length (ft)	1.5	5	12.5	20	24	63
Width (ft)	7	7	9	9	9	
Depth (ft)	1.50	1.17	0.91	0.91	0.91	
T.P.H. output	89	89	89	89	89	
Weight of clinker	1544	4002	10033	16052	19263	50893
Clinker residence time	0.5	1.3	3.4	5.4	6.5	17.2
Clinker input °F	2500	2128	1518	814	494	
Air input °F	90	90	90	90	90	
Air output °F	2140	1620	980	540	340	
Lb air (residence time)	287	1625	8040	11490	13530	34972
Clinker output °F	2128	1518	814	494	317	
Air needed (SCFM)	7337	16023	31620	28242	27297	110518

MBtu/t heat consumption 4.1

Air needed in kiln (lb/min) 4971.52 = 66121.2 SCFM

Excess cooler air (SCFM): 44397 = 40.2% of total air in

18.3   CLINKER RESIDENCE TIME

Clinker residence time in the cooler can be determined on a theoretical basis when certain factors are known. For the purpose of illustration, three different cooler operating conditions will now be considered to determine the effects of changes of these critical variables.

Residence Time at Cosntant Output.   Under these circumstances, the bed depth is directly proportional to the grate speed. Knowing the area of the cooler, the bed depth under normal operating conditions for a given kiln output rate, and the density of the clinker, the residence time can then be calculated by Eq. (18-1).

$$t = \frac{60ahy}{R} \qquad (18\text{-}1)$$

where:

t = clinker residence time in the cooler (min)
a = area of the cooler grate surrface (ft^2 or m^2)
h = bed depth (ft or m)
y = clinker density (lb/ft^3 or kg/m^3)
R = kiln output (lb/h or kg/h)

Bed Depth at Constant Speed.   At constant grate speed, the depth of the clinker bed in the cooler varies as the kiln output varies. By solving Eq. (18-1) for h,

$$h = \frac{tR}{60ay} \qquad (18\text{-}2)$$

In both examples volume in cubic feet of clinker residing in the cooler at any time is given by

$$V = \frac{tR}{60y} \qquad (18\text{-}3)$$

Residence Time at Constant Bed Depth.   As mentioned previously, it is not desirable to maintain a constant undergrate pressure under all kiln output rates. This, of course, requires a constant bed depth, which means that the clinker residence time must be changed as the kiln output changes. Eq. (18-1) is used.

An operator should have no trouble in finding solutions for the particular cooler he is operating, as the constants in the preceding equations can be substituted so that they apply to his cooler and kiln operation, and can be developed into tables similar to Tables 18.3, 18.4, 18.5, and 18.6.

The reader's attention is especially directed to Table 18.5 which clearly shows the extremes that residence time can span at kiln output rates not uncommonly obtained during upset operating conditions. It also substantiates the statement that hot clinker-discharge temperatures can become a reality if insufficient time is allowed to cool the clinker properly.

It should now be clear that proper cooling is not obtained by operating with a constant undergrate pressure (bed depth) at all times, as it is definitely wrong to assume that undergrate pressure is a function of bed depth only. If this were so, cooler components would not burn up at such an alarming rate as now encountered on many rotary kilns.

18.4   PARTICLE SIZE OF CLINKER

A critical factor in undergrate pressure reactions is the average particle size of the clinker in the cooler. The reader might be familiar with the Blaine cement surface-testing procedure in which measurement is made of the time required to pass a specified amount of air through a standard sample of cement. It takes longer for the air to pass through a fine sample than a coarse one because the finer material imposes greater resistance against the air. Exactly the same action takes place when the clinker in the cooler gets finer because of some upset kiln operating condition. A finer clinker bed imposes more resistance against air flow, undergrate pressure increases, and the fan has to use more force to push the air through this kind of a bed than through a normal one. This however is not the only problem. Individual clinker particles are lighter when the particle size diminishes. Such particles can easily be lifted into the air stream above the cooler grates because of their lighter weight.

From this it becomes clear that there are two changes in the cooler, one following the other. First the undergrate pressure increases when the clinker gets finer because the smaller particles impede air flow through the bed. Then, when the air-flow rate is increased to restore the normal flow of air through the bed, the clinker bed can become fluidized. As soon as this takes place, resistance of the bed to the flow of air decreases and the

TABLE 18.3

COOLER PARAMETERS:
Clinker residence times at constant kiln output (Grate speed varies)

Kiln output		Bed depth		Clinker weight in the cooler		Clinker volume in the cooler		Residence time
sh.t/h	m.t/h	in.	cm	lb	kg	ft³	m³	min
73	66.2	7	18	31657	14361	323.0	9.1	13
73	66.2	8	20	36180	16412	369.2	10.5	15
73	66.2	9	23	40702	18464	415.3	11.8	17
73	66.2	10	25	45225	20516	461.5	13.1	19
73	66.2	11	28	49747	22567	507.6	14.4	20
73	66.2	12	30	54269	24619	553.8	15.7	22
73	66.2	13	33	58792	26670	599.9	17.0	24
73	66.2	14	36	63314	28722	646.1	18.3	26
73	66.2	15	38	67837	30773	692.2	19.6	28
73	66.2	16	41	72359	32825	738.4	20.9	30
73	66.2	17	43	76882	34877	784.5	22.2	32
73	66.2	18	46	81404	36928	830.7	23.5	33
73	66.2	19	48	85927	38980	876.8	24.8	35

Length of cooler grates:	63.0 ft	19.2 m
Width of cooler grates:	8.8 ft	2.7 m
Grate area:	553.8 ft²	51.4 m²
Clinker density:	98.0 lb/ft³	1570.0 kg/m³

TABLE 18.4

COOLER PARAMETERS:
Clinker bed depth variation with changing kiln outputs (Grate speed constant)

Kiln output		Bed depth		Clinker weight in the cooler		Clinker volume in the cooler		Residence time
sh.t/h	m.t/h	in.	cm	lb	kg	ft³	m³	min
62	56.2	10	26	45467	20625	463.9	13.1	22
64	58.0	10	26	46933	21291	478.9	13.6	22
66	59.8	11	27	48400	21956	493.9	14.0	22
68	61.7	11	28	49867	22621	508.8	14.4	22
70	63.5	11	29	51333	23287	523.8	14.8	22
72	65.3	12	30	52800	23952	538.8	15.3	22
74	67.1	12	30	54267	24617	553.7	15.7	22
76	68.9	12	31	55733	25283	568.7	16.1	22
78	70.7	13	32	57200	25948	583.7	16.5	22
80	72.5	13	33	58667	26613	598.6	17.0	22
82	74.3	13	34	60133	27279	613.6	17.4	22
84	76.2	14	35	61600	27944	628.6	17.8	22
86	78.0	14	35	63067	28609	643.5	18.2	22

Length of cooler grates: 63.0 ft   19.2 m
Width of cooler grates: 8.8 ft   2.7 m
Grate area: 553.8 ft²   51.4 m²
Clinker density: 98.0 lb/ft³   1570.0 kg/m³

TABLE 18.5

COOLER PARAMETERS:
Residence time when grate speed is adjusted for changing kiln output rates
(Bed depth is held constant)

Kiln output		Bed depth		Clinker weight in the cooler		Clinker volume in the cooler		Residence time
sh.t/h	m.t/h	in.	cm	lb	kg	ft³	m³	min
62	56.2	12	30	54269	24619	553.8	15.7	26
64	58.0	12	30	54269	24619	553.8	15.7	25
66	59.8	12	30	54269	24619	553.8	15.7	25
68	61.7	12	30	54269	24619	553.8	15.7	24
70	63.5	12	30	54269	24619	553.8	15.7	23
72	65.3	12	30	54269	24619	553.8	15.7	23
74	67.1	12	30	54269	24619	553.8	15.7	22
76	68.9	12	30	54269	24619	553.8	15.7	21
78	70.7	12	30	54269	24619	553.8	15.7	21
80	72.5	12	30	54269	24619	553.8	15.7	20
82	74.3	12	30	54269	24619	553.8	15.7	20
84	76.2	12	30	54269	24619	553.8	15.7	19
86	78.0	12	30	54269	24619	553.8	15.7	19

Length of cooler grates: 63.0 ft 19.2 m
Width of cooler grates: 8.8 ft 2.7 m
Grate area: 553.8 ft² 51.4 m²
Clinker density: 98.0 lb/ft³ 1570.0 kg/m³

TABLE 18.6

COOLER OPERATING STUDY

(same as base case but higher bed depth, kiln output and total air into cooler the same)

		Compartment					Total
		1	2	3	4	5	
Length (ft)		1.5	5	12.5	20	24	63
Width (ft)		7	7	9	9	9	
Depth (ft)		1.83	1.33	1.03	1.03	1.03	
T.P.H. output		73	73	73	73	73	
Weight of clinker		1883	4562	11356	18169	21803	57773
Clinker residence time		0.8	1.9	4.7	7.5	9.0	23.7
Clinker input	°F	2500	2137	1537	940	741	
Air input	°F	90	90	90	90	90	
Air output	°F	2090	1500	790	350	222	
Lb air (residence time)		350	1980	9800	14000	16500	42630
Clinker output	°F	2137	1537	940	741	641	
Air needed (SCFM)		6015	14047	27930	24937	24123	97052

MBtu/t heat consumption   4.1

Air needed in kiln (lb/min)   4077.76 = 54234.3 SCFM

Excess cooler air (SCFM):   42817.7 = 44.1% of total air in

undergrate pressure will suddenly drop. A fluidized clinker bed is a highly undesirable and dangerous condition because a bed in such a state usually does not move along properly in the cooler. On a horizontal grate cooler the clinker bed when fluidized tends to remain stationary with more and more clinker building up on it. Then when sufficient weight has been acquired by the bed and it starts to move again, there is so much clinker present in the cooler that it cannot be properly cooled, and it could again choke off air flow through the bed, starting the cycle over again.

It is a fundamental rule of traveling grate cooler operation never to permit raw feed or extremely fine clinker to enter the cooler. The most modern traveling grate cooler in use today is not designed to handle such fine materials and can become overheated and damaged if called upon to do so. The best procedure in such an instance is to lower the kiln speed in order that the load in the cooler is lessened, thus giving enough time for the fine clinker to cool properly.

Another question which has to be addressed is: "How does the clinker-discharge temperature react when the kiln output remains the same but the clinker bed depth in the cooler is increased?" Such an increase in bed depth is usually the result of:

a) deliberate decrease in grate speed due to the selection of a higher *undergrate-pressure* setpoint or
b) a design change where the effective grate area of the cooler is reduced.

Naturally, this increase in bed depth (undergrate pressure) immediately imposes a higher restriction on the cooler air fans and demands sufficient fan capacity to overcome this added restriction. Consider the base case (Table 18.1). Here, 42,630 lb of air are given to the cooler in 20.9 min to cool 50,893 lb of clinker. Assume that this cooler would be operated with a higher bed depth but with the same kiln output and 42,630 lb air input. This operating condition is shown in Table 18.6 and indicates that the clinker residence time has increased from 20.9 to 23.7 min. due to the increase in residence time, the 42,630 lb of air now represents a specific air-flow rate of only 97,052 SCFM. The most important aspect of this change is that the resultant increase of the clinker-discharge temperature from 224 F to 641 F is completely unacceptable and dangerous. Here too, the solution is to increase the specific air-flow rate (SCFM) to compensate for the increase in bed depth.

18.5   OPERATION OF COOLER FANS

Now consider control of the rate of air flow in the cooler. The prime requirement in cooler control is to make sure that the air flow through the clinker bed is never restricted completely, because such a restriction leads directly to insufficient cooling of the clinker and possible damage to the cooler components. It has been pointed out that different clinker beds exert different resistances against the cooler fans. To be able to understand this important fact more clearly, the operator must have some knowledge of how a fan works under actual operating conditions.

Fig. 18.2 is a simplified illustration of an undergrate compartment with its corresponding air fan. Air-volume control for cooler fans is most commonly carried out by means of either fan outlet dampers, fan inlet vanes, or sometimes both. The fan speed is constant for any one cooler fan so it is necessary for the operator to change the position of the damper or vane to reduce or increase the volume of air moved by the fan assuming no change in static pressure on the fan.

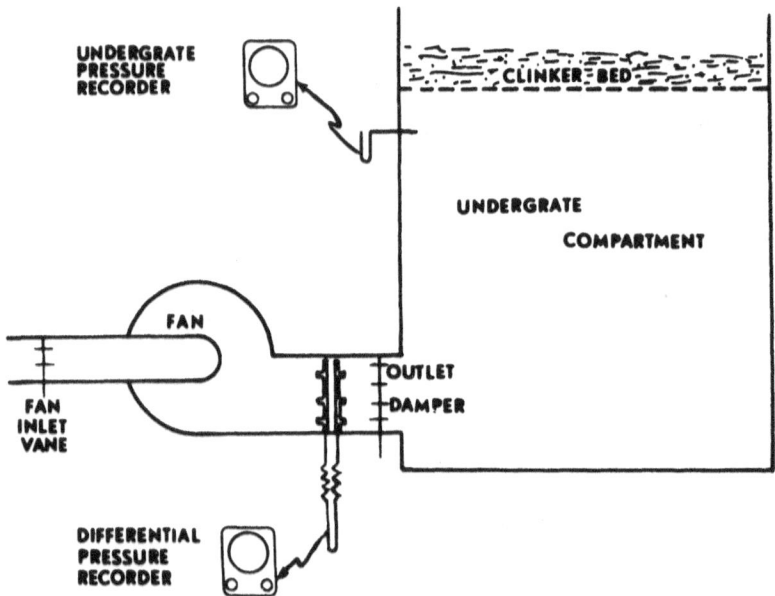

Fig. 18.2   Schematic diagram of one undergrate compartment of a traveling-grate cooler.

Fan manufacturers usually supply performance curves of individual fans for the customer. Fig. 18.3 is such a curve showing the volume of air plotted against horsepower and static pressure. In order that this discussion does not become too technical and enter fields over which an operator has no control, the horsepower curve will be disregarded and attention directed to the static pressure curve.

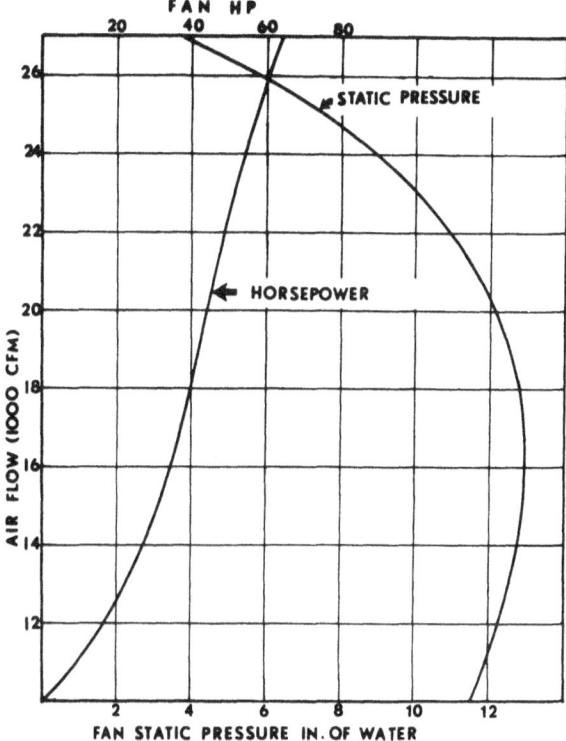

Fig. 18.3 Typical fan performance curve.

Fan static pressure is the total pressure developed by the fan, less the velocity pressure in the fan discharge duct. For all practical purposes, fan static pressure in a cooler installation is equal to the undergrate pressure, within close tolerances. Air flow is a function of static pressure and power applied to the fan. To make this discussion more meaningful, substitute undergrate pressures for equal units of fan static pressures, and restrict the range of pressures to values as they are encountered normally on rotary-kiln coolers, thus obtaining new fan performance curves as shown in Fig. 18.4. The curve on the right represents the same curve as shown in Fig. 18.3, that is, the dampers of the fan are wide open, and the air flow is the maximum obtainable for this particular fan. This curve shows clearly how the air volume decreases as the undergrate pressure increases. For example, when the undergrate pressure is 9 in. of water with the damper wide open, the volume of air moved by the fan is 24,000 CFM. If the undergrate pressure increases, perhaps as a result of increased kiln output, the volume of air progressively decreases until at 12 in. undergrate pressure the volume is only 20,000 CFM. With the fan dampers wide open, an increase in undergrate pressure causes less air to pass through the clinker bed, creating the dangerous possibility of an overheated cooler and at the same time insufficient cooling of the clinker.

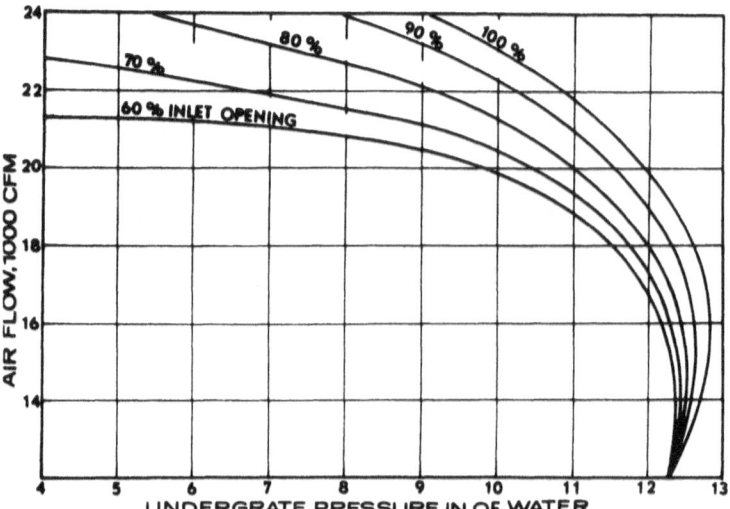

Fig. 18.4 Performance curve for a fan under different inlet vane openings. Operation is apt to be critical and unstable at pressures above 12 in. for this particular fan.

During normal and stable operating conditions, the fan dampers or vanes are never fully open. In Fig. 18.4 additional curves have been plotted showing inlet vane openings of 90, 80, 70, and 60%. Usually, a cooler fan is operated at approximately 60% opening, thus giving the operator the necessary freedom to increase air flow if a kiln upset should make it necessary. For example, assume that the cooler is operating in a stable fashion with adequate cooling of the clinker at an undergrate pressure of 7 in., fan inlet vane at 60% opening, and air volume of 21,000 CFM. For some reason, the undergrate pressure increases to 11 in. In order to obtain the same volume of air through the clinker bed, the fan inlet vane opening must be increased from 60% to 90%. If the vane were not opened, the air volume would drop to 19,000 CFM.

These examples of fan performance under actual operating conditions point out two significant facts: First, as with any butterfly-type valve or damper, maximum flow is reached when the damper is about 88% open. Any enlargement in the opening thereafter gives only a very small increase in air flow. Second, air-volume output is directly related to undergrate pressure. Therefore, the fan must have sufficient capacity to provide the necessary amount of air at the maximum anticipated undergrate pressure.

A word of caution must be introduced here. Each fan installation will have its own characteristics. The fan curves discussed in this chapter apply to one certain installation and can be considered as typical, to show the reader the relationships between pressure, output, and power requirements. Reactions and capacities of any individual installation must be computed for that particular combination of equipment. For this reason, it is advisable for an operator to examine the curves of the particular cooler fans that are under his control, and become familiar with their characteristics and capacities, so he will know the operating limits for undergrate pressure. Whenever the undergrate pressure exceeds these limits, the operator will know then that less efficient cooling is taking place and that an immediate change in kiln speed must be made to avoid damage to the cooler.

18.6 CLINKER AND AIR DISTRIBUTION IN THE COOLER

Finally, it is necessary to consider distribution of clinker and air in the cooler. For proper cooling of the clinker it is essential that the clinker is evenly spread over the width of the cooler so that the bed offers uniform resistance to the passage of air throughout its width. When the clinker passes to one side of the cooler, leaving a thinner bed on the opposite side, the air will naturally seek a passage through the bed where it offers the least resistance. Consequently, the air passes through the bed where it is least needed and little air passes where it is needed most. Formation of stalagmites (commonly referred to as "snowmen" or "candles") at the cooler inlet is the prime cause of this condition. Various devices are used to combat stalagmite formations. Some coolers have watercooled steel jackets, or watercooled clinker spreaders, and others have a special row of quench grates with their own air supply, to spread the clinker rapidly over the width of the cooler at the inlet.

For proper distribution of air in the cooler no air should freely pass from one undergrate compartment to another through large leaks or other openings. If this is allowed to take place, the air introduced, for example into the first compartment, could pass over into the second compartment when the clinker bed is thicker at the cooler inlet.

Proper clinker distribution is an acute problem in the upper region of many grate coolers. This problem can usually be overcome by:

a) narrowing the cooler width by means of installing refractory ledges and or "dead" grates along the cooler walls and

b) concurrently increasing the static-pressure capability of the air fans in these narrowed compartments.

In Chapter 12 it was briefly mentioned how important it is to apply the maximum air in these compartments where it can do the most thermal work. This means, the majority of the cooling work should be accomplished in the first and second compartments because it is here that the most efficient heat transfer will take place due to the large temperature difference between the air and the clinker. Once again, the base case in Table 18.1 is considered and it is assumed that 2000 lb more air is added into the fourth cooler compartment, i.e., at the place in the cooler where heat transfer is relatively inefficient. Table 18.7 shows that this 4.7% increase in total air input results in a clinker-discharge-temperature drop of only 13 F (from 224 F to 211 F). It is assumed that all compartment air flows are readjusted to take advantage of the aforementioned thermo-dynamic principle. In Table 18.8 air has been added to the second and third compartments while it has been reduced to the fourth and fifth compartments. The end result is that the total air input into the cooler has been

TABLE 18.7

COOLER OPERATING STUDY

(same as base case but 2000 lb more air into fourth compartment)

		Compartment					Total
		1	2	3	4	5	
Length (ft)		1.5	5	12.5	20	24	63
Width (ft)		7	7	9	9	9	
Depth (ft)		1.50	1.17	0.91	0.91	0.91	
T.P.H. output		73	73	73	73	73	
Weight of clinker		1544	4013	10033	16052	19263	50905
Clinker residence time		0.6	1.6	4.1	6.6	7.9	20.9
Clinker input	°F	2500	2057	1365	642	355	
Air input	°F	90	90	90	90	90	
Air output	°F	2090	1520	840	420	240	
Lb air (residence time)		350	1980	9800	14000	18500	44630
Clinker output	°F	2057	1365	642	355	211	
Air needed (SCFM)		7339	15968	31613	28225	30614	113758

MBtu/t heat consumption   4.1
Air needed in kiln (lb/min)   4077.76 = 54234.3 SCFM
Excess cooler air (SCFM):   59524.2 = 52.3% of total air in

TABLE 18.8

COOLER OPERATING STUDY

(same as base case but adding more air in second and third, and reducing air in fourth compartment)

		Compartment					
		1	2	3	4	5	Total
Length (ft)		1.5	5	12.5	20	24	63
Width (ft)		7	7	9	9	9	
Depth (ft)		1.50	1.17	0.91	0.91	0.91	
T.P.H. output		73	73	73	73	73	
Weight of clinker		1544	4013	10033	16052	19263	50905
Clinker residence time		0.6	1.6	4.1	6.6	7.9	20.9
Clinker input	°F	2500	2057	1221	498	300	
Air input	°F	90	90	90	90	90	
Air output	°F	2090	1500	790	350	222	
Lb air (residence time)		350	2425	10500	12300	11000	36575
Clinker output	°F	2057	1221	498	300	224	
Air needed (SCFM)		7339	19556	33871	24798	18203	103767

MBtu/t heat consumption   4.1
Air needed in kiln (lb/min)   4077.76   = 54234.3 SCFM
Excess cooler air (SCFM):   49532.3   = 47.7% of total air in

reduced by 15% but the clinker-discharge temperature remains the same at 224 F. Once again, it is important for an operator to remember this principle of cooler-air application as this will lead to efficient cooler operation and possibly to less failures and damages to cooler components.

Undergrate pressures are usually set so that the highest pressure is found in the first compartment and the lowest in the last compartment. Clinker temperatures above the grates, as pointed out earlier, influence the undergrate pressure. This step-by-step lowering of the undergrate pressures from one compartment to another is a natural result of the succeeding lower temperaures of the clinker as it travels down the cooler.

18.7 SECONDARY AIR-TEMPERATURE CONTROL

Secondary air temperature has a direct influence on the geometry of the flame and the point of ignition of the fuel, consequently irregular secondary air temperatures can cause irregular flame characteristics which in turn can cause a shifting of the burning zone. Stable kiln conditions are practically impossible as long as the secondary air temperature is not held constant, within a tolerance of ± 100 F (40 C). The following discussion applies to controls to be exercised when the kiln is operating at normal speed and is producing well-burned clinker.

The kiln should be operated with the secondary air temperature as high as possible because the maximum amount of heat is then recovered from the clinker, thus improving fuel efficiency, as less fuel is required to raise the temperature of the air entering the kiln. This condition maximizes kiln capacity. A second advantage of high secondary air temperature is the favorable influence on the objective of burning the clinker close to the nose (front) of the kiln. However, there are practical limits for secondary air temperatures. A temperature that is too high can result in overheating in the kiln nose and the burner hood as well as in the cooler. It is advisable to operate the kiln with the secondary air temperature slightly below the maximum allowable in order to protect the bed grates and cooler refractory from damage.

The two factors having the greatest influence on secondary air temperature are speed of the bed grates in relation to the volume and temperature of air introduced into the cooler to cool the clinker, and temperature and size of clinker discharging from the kiln into the cooler. Under stable kiln conditions, the secondary air temperature is controlled by the speed of the cooler grates which merely means that the depth of the clinker bed in the cooler is the controlling factor. Thus, an increase in grate speed (lessened bed depth), other conditions remaining unchanged, results in a lower secondary air temperature, and a slower grate speed (thicker bed depth) causes an increase in temperature. It is important to remember, however, that secondary air temperature control is not merely a matter of speeding up or slowing down movement of the bed grate. As mentioned earlier, there are several factors to be considered before deciding which adjustment will give the desired results.

Because the secondary air temperature is controlled mainly by the bed-grate speed and the volume and temperature of the air in the cooler, as well as the temperature and size of the clinker, this control goes hand in hand with undergrate-pressure control, as any change in these two variables will also change the undergrate pressure. This then will considerably limit the extent to which secondary air temperature can be controlled.

Now consider for example, an upset kiln condition in which the greatest part of the material in the cooler is in the form of very small-sized nodules, or even worse, in the form of dust. In a situation like this the operator will first reduce the kiln speed, which reduces the amount of clinker and consequently lowers the secondary air temperature (see Table 18.9). If an attempt is made to hold the secondary air temperature within a 100-degree tolerance, it would be necessary to slow down the bed-grate speed to such an extent that it would choke off the free passage of air through the bed. The clinker would thus not be properly cooled and would probably be red hot on leaving the cooler, causing considerable damage to the clinker transport equipment. The proper adjustment in this case would be to slow down the bed grates only to such an extent that the normal bed depth can be maintained. It is also extremely important that the air flow into the cooler be increased in order that the clinker can be properly cooled. The operator should never attempt to hold the secondary air temperature at its normal level when the kiln has been slowed down because of an upset.

The other factor to be considered in secondary air-temperature control is the temperature of the clinker as it discharges from the kiln. By changing the character of the flame, the burning zone can be shifted closer to the kiln front thus raising the clinker-discharge temperature and consequently the secondary air temperature, or the burning zone can be shifted further back in the kiln, reducing the secondary air temperature. These actions are summarized in Table 18.9.

TABLE 18.9

SECONDARY AIR-TEMPERATURE CONTROLS

To reduce secondary air temperature:	To raise secondary air temperature:
Increase cooler grate speed.	Decrease cooler grate speed.
Decrease clinker bed depth in cooler.	Increase clinker bed depth in cooler.
Decrease size of clinker particles.	Increase size of clinker particles.
Decrease amount of clinker.	Increase amount of clinker.
Burn further back in the kiln.	Burn closer to the nose (front) of the kiln.

Earlier in this chapter it was pointed out that more air must always be given to the cooler than what is needed for combustion in the kiln. Hence, a certain amount of excess air has to be vented to the atmosphere or used for the drying and grinding of raw materials and/or coal. On precalciner kilns, this excess air is diverted to the flash calciner by means of the tertiary air duct. By inference there must therefore be an imaginary dividing line in the cooler wherein the air in the hotter region goes to the kiln and the one in the colder part moves toward the cooler stack. Simple calculations on the operating conditions explained in Tables 18.1 and 18.8 would show that more efficient application of the air in the first two compartments would result in a shift of this imaginary line toward the upper end of the cooler, higher secondary air, and lower cooler-stack air temperatures.

Design limitations, either in cooler size or fan capacities, often lead to excessively high clinker-discharge temperatures regardless of how well the air distribution is controlled or the cooler mechanically maintained. Such kilns are usually capable of producing more clinker but the cooler acts as the bottleneck toward these higher output rates. Short of installing an after-cooler (such as the well-known G-cooler) or major modifications in cooler design, there is not much one can do to overcome these limitations. Some plants use water-spray cooling of the clinker within the last cooler compartment (usually leading to operating problems with the dust collector) or make use of reciprocating clinker "skips" combined with water sprays after the clinker discharges from the cooler. These solutions must be viewed as only temporary as most of these create additional operating problems.

18.8　HOOD-DRAFT (PRESSURE) CONTROL

A definitive volume of air is drawn into the kiln by the I.D. fan and a given amount of air is forced into the cooler by the cooler fans. Years ago, on wet-process kilns with their high specific-heat consumption and relatively low kiln output, these two flows were nearly equal. Hence, only small amounts of excess air had to be vented to the atmosphere by means of the cooler stack. Some of these kilns were equipped with so-called closed-circuit cooler-air systems (recycling excess air to the upper cooler compartment fans) and successfully eliminated all excess air. However, as kilns became more efficient in heat consumption and clinker output rates, excess cooler-air volumes started to increase and in many cases this led to difficulties in hood-draft control. All too often the damper in the cooler stack which regulates the hood draft is found to be in the fully open position thus eliminating any effective control of the hood draft.

Hood-draft control is simply a regulation of the amount of excess air which escapes through the cooler chimney, so that when the hood pressure is too high, the damper must be opened, and when the hood pressure is too low, the damper must be closed. This does not mean that the damper has to be fully closed or fully opened, but instead small adjustments in the damper position can be made to give the desired results.

As mentioned, hood draft is governed mainly by:

a) the I.D. fan speed and
b) the volume of air given to the cooler fans.

Assuming that other factors remain constant, then an increase in I.D. fan speed results in a lower hood pressure, and a reduction in I.D. fan speed results in a higher hood pressure. Similarly, increasing the amount of air forced into the cooler results in higher hood pressure.

The hood pressure can be either negative or positive. A negative pressure indicates that the hood is under a vacuum, and a positive hood pressure means that the hood is pressurized. The basic rule in hood-pressure control is never to operate a kiln with positive hood pressure because this results in troublesome kiln operating conditions. Fine clinker particles are blown through the nose-ring seal thus causing the seal to wear out prematurely, and dust emission in the hood area can make viewing of the burning zone by the operator unpleasant and unsafe. On rotary kilns equipped with optical pyrometers and television cameras for burning-zone control, positive hood pressures could lead to damage to this equipment from the flying hot particles. Formation of rings and of stalagmites ("snowmen") in the cooler inlet can be attributed to positive hood pressures on some kilns. The importance of operating a kiln with negative hood pressure is obvious.

There is one situation in which the operator has to take exception to the above rule: Whenever prevailing high cooler temperatures could cause damage to the cooler components, it is necessary to introduce a sufficient amount of air into the cooler to lower the temperature and overcome the dangerous situation. The first corrective action in such an instance would be to slow down the kiln speed to lessen the load in the cooler. Then if dangerous overheated conditions still prevail and the damper at the chimney is already wide open, one has no other choice but to introduce an added volume of air into the cooler. However, the operator should return the hood pressure to negative again at the earliest possible time as soon as the situation has been brought under control.

The hood pressure is usually automatically controlled. The controller receives an input signal from the hood-pressure measuring instrument and sends an output signal to the damper at the cooler chimney. The operator adjusts the setpoint on the controller to the desired hood pressure (generally between −0.07 and −0.03 in. of water). Any time the hood pressure deviates from this setpoint, the chimney damper will then be automatically adjusted by the controller.

19.

Kiln Exit-Gas Temperature Control

The technique of burning a kiln from the rear was briefly mentioned in Chapter 14. In this technique, equal or more consideration is given to the so-called kiln back end (temperature and oxygen content) than to the burning zone because proper and stable back-end conditions lead to stable burning-zone conditions. Full control must also consider the pressure or draft conditions at the rear of the kiln, as the draft pressure is an indication of the presence of several irregularities that affect operation and control of the kiln.

19.1   BACK-END TEMPERATURE

Whenever the term "back-end temperature" is used it refers to the exit-gas temperature in wet and dry kilns without chains and to the intermediate gas temperature in kilns equipped with chains. For preheater and pre-calciner kilns, back-end temperature herein refers to the rotary-kiln exit-gas temperature, i.e, the gas conditions at the inlet to the lowest stage of the preheater cyclones.

A tight back-end temperature control is the key to a successful, stable kiln operation. There is sound reasoning behind the procedure of controlling the back-end temperature within a narrow range: kiln feed while traveling through the kiln undergoes physical and chemical changes all of which are important and cannot be overlooked if the kiln is to operate in as stable a condition as possible. The most crucial change affecting the feed takes place in the calcining zone. Proper burning of the feed in the burning zone cannot take place unless the feed is completley calcined before it enters the burning zone. In other words, clinkerizing will not proceed until all the carbon dioxide (CO_2) has been driven off the feed. It should now be clear that a certain heat gradient has to exist throughout the kiln in order that drying and calcination proceeds in the desired manner. There must be means of controlling calcination and drying of the feed before it enters the burning zone. Back-end temperature control is the means by which the operator keeps control over feed preparation behind the burning zone.

The fact that the operator controls the back-end gas temperature instead of the feed temperature (whenever both are measured) raises the question of whether the feed temperature in the kiln is more important than the gas temperaure. This of course is true, but the important fact is that the gas temperature is easier to control than the temperature of the solids.

On dry- and wet-process kilns, the solids temperature does not react as fast as the gas temperature whenever a change in the control settings has been made. By controlling the gas temperature, the operator can observe a change minutes after an adjustment has been made, but considerable time could pass before a reaction in the temperature of the solids would be observed. Nevertheless, it would be wrong to assume that the solids temperature at the back end of the kiln is not important. This temperature does reveal considerable information in regard to the heat transfer taking place in the chains.

Ideal Conditions. For any kiln speed and feed rate on a kiln, there is an ideal back-end temperature that will ensure proper preparation of the feed. If too much heat exists toward the rear of the kiln, the feed will unnecessarily undergo early completion of calcination and clinker formation will start further back into the kiln. Aside from shifting the burning zone further back, too much heat to the rear of the kiln represents poor operating efficiency and results in a waste of fuel. In the opposite direction, not enough heat to the rear of the kiln could cause the undesirable condition in which feed is not completely calcined when it enters the burning zone, rendering the burnability extremely difficult and probably leading to an upset in the operation.

The back-end temperature is governed by several factors. Of these, the I.D. fan speed and the fuel rate are the usual causes for changes in back-end temperature. Whenever one variable is changed and all others remain constant then *higher* back-end temperature will result if there is an increase in I.D. fan speed, an increase in fuel rate, or a decrease in feed rate.

Conversely, *lower* back-end temperature results from a decrease in I.D. fan speed, a decrease in fuel rate, or an increase in feed rate.

Changes in feed rate usually do not happen in a mechanically-sound rotary kiln if the kiln speed remains constant. Nevertheless, from time to time conditions will arise in which the feed loading of the kiln changes due to mechanical failure of the feeding system, feed holdback in the heat exchangers (chains, preheater cyclones), or most commonly, alteration of the kiln speed. This in turn will have a direct influence on the back-end temperature.

To summarize, changes in the following variables can lead to changes in the back-end temperature:

a)   I.D. fan speed (kiln-exit draft)
b)   fuel rate in the burning zone
c)   feed rate
d)   kiln speed

Clearly, this again points out an important aspect of kiln control, namely that changes in fuel rate, kiln speed, and/or I.D. fan speed cause changes at both ends of the kiln (burning zone and back end). Hence, if one changes any of the above-mentioned kiln-control variables, one has to consider the effect this change might have throughout the entire kiln. For example, if the back-end temperature is low one can not take a simple approach and just increase the I.D. fan speed without considering the effect this adjustment has on the burning-zone temperature and oxygen content in the exit gases.

19.2   BACK-END DRAFT CONTROL

Back-end draft, or any kind of draft measurement on the kiln system, is measured by means of a sensing tube inserted into the gas stream (see Chapter 12). For back-end draft, one measures predominantly the local static pressure. Since the kiln operates under induced draft by the large fan at the kiln back end (hence the name I.D. fan) this local pressure is negative, i.e., the suction is indicated by the instrument. This is important to remember because it explains the reasons why this instrument reading either increases or decreases under certain kiln conditions. If there is a restriction downstream (further down the kiln) from this measuring point, this "pressure" increases because there is more suction. However, if the restriction is between the I.D. fan and the point of measurement (i.e., upstream) this suction decreases.

a) Long dry- and wet-process kilns.

In the burning technique previously described, the operator uses the back-end draft pressure recorder as an indicator of ring formation, slabbing of rings, feed holdbacks in the chain section, irregularities in the I.D. fan performance, or for the detection of possible air leakage in the system between the kiln rear and the I.D. fan. The back-end draft should not be used for regulating the temperature profile in the kiln because too many factors can cause a change in this draft. In order to achieve satisfactory kiln operation, the main objectives should be a constant back-end temperature, an oxygen range of 0.7–1.5%, and a secondary air temperature as nearly constant as possible.

Some plants have attempted to control I.D. fan speed in closed-loop automatic-control mode by using the back-end draft as the sole input signal to the controller. In other words, the I.D. fan would only change if there would be a change in back-end draft. Such efforts fail in most instances for the reason that the I.D. fan is not able to provide the kiln with the constant air flow rate that was wrongfully assumed. This also ignores the temperature profile of the kiln, the rings, or other areas where buildup might take place. The same applies to attempts that have been made to control the I.D. fan speed based on an input signal from the percent oxygen in the exit gases. Constant oxygen levels, in theory, assures a constant fuel-to-air ratio. This is a great concept for the internal combustion engines but has little value for rotary-kiln control. Both these concepts might have some limited application possibilities during stable kiln operation but are definitely counterproductive during upset conditions. The truism can again be repeated: "Kiln burning is not so much a task of maintaining stable operation but rather a specialized skill of knowing how to prevent upsets and how to stabilize a kiln after an upset." Kiln operators are being measured by these skills with the same criteria applying to the control program that is stored in the memory in the computer.

Changes in the I.D. fan speed are the most likely causes of changes in the back-end draft. This is only natural because with every alteration in the I.D. fan speed the velocity of the gases passing through the kiln changes. Hence, whenever the operator detects a change in the back-end draft his first question should be: "Was the I.D. fan speed altered just prior to the time this change in the draft was observed?" If the answer is negative, then he can most likely find the reason in one of the following:

1. an inspection door between the kiln and the I.D. fan has been opened or closed;
2. the fuel chamber door has been opened or closed (as for example when the dust from this chamber has to be emptied out into a truck); or
3. the cold air breed-in damper position in the kiln rear has been changed (wherever such a device is in existence due to I.D. fan inlet temperature restrictions).

If any one of the above-mentioned doors or dampers is opened, the back-end draft will decrease because atmospheric air is being pulled into the system, reducing the amount of gases pulled through the kiln itself. On the other hand, closing of any of the above doors will cause the draft to increase. Depending on the size of the door or the damper, changes in the draft can be either very small or so large that they could lead the kiln into an upset if not counteracted at once.

Losing kiln draft in any of the ways described will lower the oxygen content in the flue gases and will also lower the back-end temperature. Under such conditions, it is advisable to adjust the I.D. fan speed to return the kiln draft to its normal level, which should also ensure a return to the normal oxygen and back-end temperature. This illustrates an important usage of the back-end draft recorder, as it indicates immediately any change in the gas velocity inside the kiln, which under stable kiln conditions should be fairly constant.

Any workman who has to open a door at the back end of the kiln must always report his intentions to the operator first, obtaining the operator's permission before making any changes. It certainly does not help the operator any if someone changes the kiln draft without his knowledge, especially if the kiln is already on the brink of an upset.

Now assume that there has been a drastic drop in the kiln draft but there has not been a change in fan speed, nor have any doors or dampers been opened. In this case the origin of the change must be sought in the kiln itself. Whenever the I.D. fan speed is constant and an obstruction, such as a ring in the kiln, is willfully or accidently removed, there will be an immediate drop in the back-end draft. On the other hand, if a ring is in the process of forming, the back-end draft will increase progressively as long as the ring keeps on building up. This is another important reason that justifies a close watch of the back-end draft. Ring formation or droppage can be easily and promptly detected by evaluation of the draft recording chart as shown in Fig. 19.1. This is especially helpful in case a ring has broken from behind the burning zone or from the chain section, giving the operator time to adjust the kiln settings for the expected heavier load which will arrive in the burning zone.

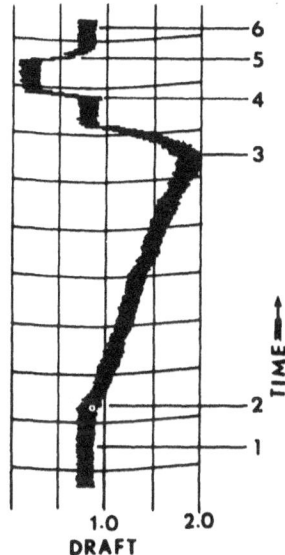

Fig. 19.1   A portion of a back-end draft recorder chart, showing: 1, normal draft; 2, a gradual and steady increase in draft pressure indicates that a ring is forming; 3, the ring is starting to disintegrate and pressure returns to normal when the ring breaks loose; 4, a sudden drop in pressure indicates that a door or louvre has been opened; 5, the door or louvre has been closed and the draft pressure returns to normal at 6.

b) Preheater and Precalciner Kilns.

In contrast to conventional long kilns, operators of preheater and precalciner kilns have the added responsibility of closely monitoring the draft conditions in each preheater vessel and, in the case of the precalciner kiln, the draft in the tertiary air duct also. The so-called pressure drop across a preheater stage is important because it gives the operator an indication of buildup problems within the vessel or discharge chutes. Whereas a ring might build up relatively slowly within a wet-process kiln over a period of a few days, an alkali-sulfate or calcium-sulfate buildup in the feed chute or lower preheater stage can progress very rapidly and, consequently, can produce large changes in draft in a short period of time.

20.

Feed-Rate Control

Most modern rotary kilns have feed-metering devices which, being synchronized with the kiln speed, maintain a uniform depth of feed bed throughout the entire range of possible kiln speeds. The cross-sectional loading of the kiln remains theoretically the same regardless of speed of rotation, as long as the feed-to-kiln speed ratio is not changed.

An element of feed rate that has to be considered is the amount of dust which is returned into the kiln, a condition that is overlooked in many cement plants. In extreme cases, dust is returned to the kiln at the same rate as it is collected in the dust-collecting units. This means that for any given time period there could be very little, and at other times there could be a large amount of dust returned to the kiln. Although the feed rate is synchronized with the kiln speed, the irregular dust return will cause an uneven loading in the kiln. This problem can be partly overcome by installing a surge bin which to some extent equalizes large variations in the dust-collecting rates. This procedure still does not feed the kiln in the same fashion as the kiln feed because the dust return is not synchronized with the kiln feed. This being the case, the operator should, if he has the means available, make an attempt to synchronize the dust rate manually with the kiln speed whenever the kiln speed is changed. If the kiln speed is decreased, less dust should be returned to the kiln. Conversely, more dust should be returned to the kiln whenever the kiln speed is increased.

Now assume that this problem does not exist and raw feed as well as dust is fed to the kiln at a synchronized rate. One could therefore conclude that the kiln would be evenly loaded from the inlet to the outlet. If this held true at all times, operation of a rotary kiln would be greatly simplified. This, however, is not the case. Feed can flush at great speed into the burning zone because of erratic calcining conditions, or feed can be retained in the chain section, or in the rear of the burning zone as a result of ring formation. When the ring breaks loose (slabbing), there will be a sudden surge of feed into the burning zone to be clinkerized. Any ring in any part of the kiln will act as a dam obstructing the uniform advancement of the feed through the kiln.

It is important that the kiln operator keep a close watch over all instruments registering feed loading or feed advancement in the kiln, so that any abnormalities can be noted as they develop. As with all other control functions, early detection is of prime importance because the earlier any irregularity is detected and counteracted, the smaller the chance of leading the kiln into a serious upset, with the resulting damage to the equipment. Chains, coating, and even refractories have been lost (melted) because of overheated conditions resulting from feed shortages and negligence on the part of the operator. Uniform feeding of the kiln is the first step toward stable kiln conditions.

When a kiln operates for a prolonged period of time in a stable fashion, there inevitably arises the question of whether the production rate of the kiln can or should be increased. Unfortunately, the answer to this question is often dictated by prevailing clinker inventories and existing commitments to cement sales. These conditions often impose higher output demands on the kiln at the expense of operating stability and ease of control. However, despite their common objections kiln operators must learn to cope with these factors.

It should not be assumed that a kiln may be increasingly "force-fed" with feed until serious and dangerous out-of-control conditions are reached. There is also no reason to belive that it is advantageous to operate a kiln continuously with output rates that are considerably below the rates that are safely attainable. Each plant should have a set of guidelines that specify for the operators when kiln production rate should be changed. Decisions for feed-rate changes should never be based on spur-of-the-moment beliefs but should instead be well thought out and planned in advance. In some plants, the decision for a kiln output change is made solely by production managers or by the kiln operator. Chances for successfully achieving higher output rates are better when both the operator and the production manager, through dialogue, agree to a warranted feed-rate increase. In short, it is good practice to obtain the input from both parties before a final decision is made.

Based on the author's operating experience, a kiln is "ready" to produce more clinker when each and every one of the following requirements are met:

1. All related kiln and grinding equipment is able to handle the increase in feed rate and output rate.
2. The kiln has operated in a stable fashion for at least 4 h prior to the intended feed increase. Burning-zone and back-end temperatures have undergone very little change.
3. The *oxygen* reading, during the previous 4 h, has been consistently above 1.3 percent.
4. The *black feed* in the burning zone, during the previous 4 h, has consistently been located behind or directly under the end of the flame.

When these four basic requirements are all met, the kiln in essence is indicating that more production could be achieved without upsetting the delicate stability of its operation.

Naturally, any feed increase should be done in a series of small steps instead of all at once. After each measured increase one whould wait for at least another 4 h to make sure that the above conditions are met before another increase is made. It is also important not to make this feed increase less than 2 h before the shift change so as not to create difficult conditions for the next kiln operator.

Regardless of whether more or less output is required, it is necessary to evaluate what effect the proposed change will have on the overall prospects of kiln conditions, fuel efficiency, clinker size, burnability, and adjustments that may have to be made to fuel rate, I.D. fan speed, kiln speed, and other variables. In general, the production rate can be changed either by maintianing the same feed-to-kiln-speed ratio and changing the kiln speed, or by continuing the same kiln speed and changing the feed ratio. Reactions in the burning zone are different from those in the back end, depending on which of the two procedures is used to bring about the change in kiln output rate.

A chart similar to Fig. 20.1, which is a portion of a chart that was designed for one particular kiln and is applicable to that kiln only, will be of value to the operator in making these adjustments. Such a chart can be made for any kiln. First, determine the maximum speed at which the kiln can be operated. Then determine the maximum feed rate in tons per hour

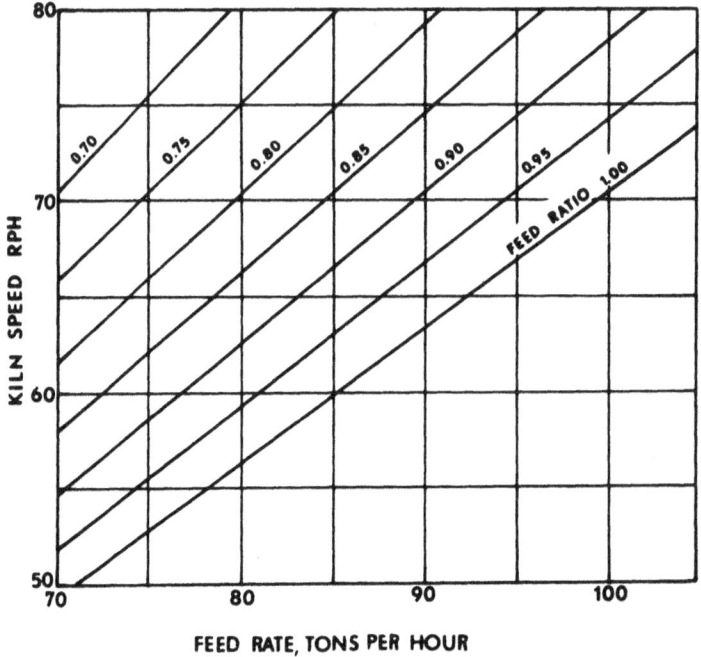

Fig. 20.1 For any operating kiln, a chart can be prepared showing the combinations of kiln speed and feed-to-kiln speed ratio that will give any selected production rate. For example, at a kiln speed of 65 rph, a feed ratio of 0.80 gives a feed rate of 74 tons per hour.

for this speed. (This information is available in the engineering data for the kiln under consideration.) The "feed ratio" on the controller is now arbitrarily set at 1.00 for this speed and feed rate. The relationship is expressed by the equation:

$$\frac{S_{max}}{S_2} = \frac{W_{max}r_2}{W_2} \tag{20-1}$$

in which

S_{max} = maximum kiln speed in rph
S_2 = required kiln speed
W_{max} = maximum feed rate in tons/h at maximum kiln speed and feed

ratio of 1.00
W_2 = required feed rate
r_2 = required feed ratio.
This equation can be solved for each of the variables:

$$r_2 = \frac{S_{max}W_2}{S_2 W_{max}} \tag{20-2}$$

$$S_2 = \frac{S_{max}W_2}{W_{max}r_2} \tag{20-3}$$

$$W_2 = \frac{S_2 W_{max}r_2}{S_{max}} \tag{20-4}$$

Examples: Assume for a certain kiln that
$S_{max} = 80$
$W_{max} = 113.6$
Determine (a) the feed ratio controller setting when
$S_2 = 75$ and
$W_2 = 100$
Using Eq. (20-2)

$$r_2 = \frac{80 \times 100}{75 \times 113.6} = 0.94$$

Determine (b) the kiln speed required when
$S_2 = 95$
$r_2 = 0.90$
Using Eq. (20-3)

$$S_2 = \frac{80 \times 95}{113.6 \times 0.90} = 74$$

It is necessary to compute only sufficient values so a family of curves can be drawn, as in Fig. 20.1. Future operating adjustments can then be made from the chart.

If the feed ratio is changed but kiln speed is held constant, the burning-zone conditions remain unchanged for 2–3 h, until the different feed rate arrives in this zone. After a short time, however, the back-end temperature starts to fall off making it necessary to adjust the I.D. fan speed and the fuel rate.

If the feed ratio is held constant and kiln speed is changed, there will be an immediate reaction in the burning zone and the back end (both cooling down) that demands an adjustment in I.D. fan speed and fuel rate almost simultaneously with the changing of the kiln speed.

Neither of the two procedures should be carried out hastily. The smaller the changes, the less the danger of leading the kiln into an upset. Resting periods of 4–8 h should be held between a series of feed-rate or kiln-speed changes in order to make sure that operation of the kiln remains stable. The above holds true only for production increases or decreases; this by no means applies to a slowdown period when the kiln has been in an upset.

There are four fundamental rules governing feed rate and feed advancement that the operator must remember:

1. Never operate a wet-process kiln longer than 10 min, or a dry-process kiln longer than 15 min, without any feed entering the kiln.
2. Never permit the back-end gas temperature to exceed the maximum allowable limit.
3. Never increase the output (or feed rate) at the expense of stable operating conditions.
4. Never delay a necessary counteractive adjustment until the change in feed-bed depth is under the flame whenever you know in advance that such a change will take place.

In the following three examples, an illustration is given of the effects of kiln-speed and feed-ratio changes on kiln-feed residence and heat-"soaking" times behind the burning zone. The data given in these tables apply to a given long dry-process kiln. Similar tables can be developed for any kiln types by using the appropriate dimensions.

Table 20.1, for example, shows a feed-ratio setting of 0.85, a clinker output rate of 88 t/h (feed rate 136 t/h) at a kiln speed of 80 rph. At this kiln speed the feed takes 2.3 h to reach the burning zone and 59.2 tons of feed is preheated and calcined each hour behind the burning zone.

If the feed ratio were changed to 1.00 (see Table 20.2) this kiln would have to operate at a speed of 68 rph to obtain the same output of 88 t/h. It is of interest to the reader to observe that the residence time of the feed behind the burning zone is now 2.7 h and only 50.3 tons of feed have to be preheated and calcined within each hour. The third example (Table 20.3) shows a more pronounced change at a feed-ratio setting of 1.24.

Here, kiln speed 55 will produce approximately the same output but the residence time is now 3.4 h and only 40 tons of feed have to be heat-treated behind the burning zone each hour.

TABLE 20.1

KILN PERFORMANCE PARAMETERS
AT VARIOUS KILN SPEEDS
(Feed ratio 0.85)

	Travel time		Feed rate		Feed exposed time
	Factor: 11035		Max. feed 200		Only preheat and
	Diameter 14.05 ft		Max. kspeed 100		calcining zone
	Slope 0.031 ft/ft		Feed ratio 0.85		
	Length 425 ft*		Factor 1.7		
	*without burning zone				

Kiln speed rph	Travel time		Feed rate tph	Clinker output tph	Feed heated tph
	min	h			
86	128	2.1	146	94	68.4
83	133	2.2	141	91	63.7
80	138	2.3	136	88	59.2
77	143	2.4	131	84	54.8
74	149	2.5	126	81	50.6
71	155	2.6	121	78	46.6
68	162	2.7	116	75	42.7
65	170	2.8	111	71	39.1
62	178	3.0	105	68	35.5
59	187	3.1	100	65	32.2
56	197	3.3	95	61	29.0
53	208	3.5	90	58	26.0
50	221	3.7	85	55	23.1
47	235	3.9	80	52	20.4
44	251	4.2	75	48	17.9
41	269	4.5	70	45	15.5
38	290	4.8	65	42	13.3
35	315	5.3	60	38	11.3
32	345	5.7	54	35	9.5
29	381	6.3	49	32	7.8
26	424	7.1	44	29	6.2
23	480	8.0	39	25	4.9
20	552	9.2	34	22	3.7

TABLE 20.2

KILN PERFORMANCE PARAMETERS
AT VARIOUS KILN SPEEDS
(Feed ratio 1.00)

	Travel time		Feed rate		Feed exposed time
	Factor: 11035		Max. feed	200	Only preheat and
	Diameter	14.05 ft	Max. kspeed	100	calcining zone
	Slope	0.031 ft/ft	Feed ratio	1	
	Length	425 ft*	Factor	2	
	*without burning zone				

Kiln speed rph	Travel time		Feed rate tph	Clinker output tph	Feed heated tph
	min	h			
86	128	2.1	172	111	80.4
83	133	2.2	166	107	74.9
80	138	2.3	160	103	69.6
77	143	2.4	154	99	64.5
74	149	2.5	148	95	59.5
71	155	2.6	142	92	54.8
68	162	2.7	136	88	50.3
65	170	2.8	130	84	45.9
62	178	3.0	124	80	41.8
59	187	3.1	118	76	37.9
56	197	3.3	112	72	34.1
53	208	3.5	106	68	30.5
50	221	3.7	100	65	27.2
47	235	3.9	94	61	24.0
44	251	4.2	88	57	21.1
41	269	4.5	82	53	18.3
38	290	4.8	76	49	15.7
35	315	5.3	70	45	13.3
32	345	5.7	64	41	11.1
29	381	6.3	58	37	9.1
26	424	7.1	52	34	7.4
23	480	8.0	46	30	5.8
20	552	9.2	40	26	4.3

TABLE 20.3

KILN PERFORMANCE PARAMETERS
AT VARIOUS KILN SPEEDS
(Feed ratio 1.24)

Travel time			Feed rate		Feed exposed time
Factor: 11035			Max. feed 200		Only preheat and
Diameter 14.05 ft			Max. kspeed 100		calcining zone
Slope 0.031 ft/ft			Feed ratio 1.24		
Length 425 ft*			Factor 2.48		
*without burning zone					

Kiln speed rph	Travel time		Feed rate tph	Clinker output tph	Feed heated tph
	min	h			
86	128	2.1	213	138	99.7
83	133	2.2	206	133	92.9
80	138	2.3	198	128	86.3
77	143	2.4	191	123	79.9
74	149	2.5	184	118	73.8
71	155	2.6	176	114	68.0
68	162	2.7	169	109	62.4
65	170	2.8	161	104	57.0
62	178	3.0	154	99	51.8
59	187	3.1	146	94	46.9
56	197	3.3	139	90	42.3
53	208	3.5	131	85	37.9
50	221	3.7	124	80	33.7
47	235	3.9	117	75	29.8
44	251	4.2	109	70	26.1
41	269	4.5	102	66	22.7
38	290	4.8	94	61	19.5
35	315	5.3	87	56	16.5
32	345	5.7	79	51	13.8
29	381	6.3	72	46	11.3
26	424	7.1	64	42	9.1
23	480	8.0	57	37	7.1
20	552	9.2	50	32	5.4

21.

Kiln Starts and Shutdowns

Until now, only individual control functions applying to steady-state kiln operations have been discussed. This chapter deals with the fundamentals of bringing a kiln on line, shutting a kiln down, and optimizing the kiln during upset conditions to lead the kiln back to stable operation after a major operating disturbance (upset).

21.1 KILN STARTS

Starting a kiln from cold state, such as after major refractory repairs, demands special consideration toward the following items:

a) A complete system check to make sure all main and auxiliary equipment are ready for operation
b) Drying of refractory lining
c) Expansion of refractory lining, kiln shell, and tires
d) Timing of the moment when the first feed is given to the kiln
e) Timing of the burning-zone refractory temperature so that coating is immediately formed and clinker produced at the moment of arrival of the first feed to the burning zone.

These tasks are made easier when an empty kiln is being started, i.e., a kiln that contains very little or no feed prior to the start-up. In these instances, the kiln-start sequences and measured temperature increases can be predetermined and preprogrammed. This is not the case for a kiln that had been shut down under load and has cooled down sufficiently to require a long kiln warm-up to bring it back to operating temperature. In such instances, clinker will have to be produced within a short period of time after the main drive has been activated, thus usually creating too fast a warm-up period with possible damaging effects on the kiln shell and refractory lining. It is important to examine the special considerations of kiln start-up more closely.

a) Complete Systems Check. The most useful and essential means of ascertaining complete system readiness is a written checklist wherein operating and maintenance personnel can indicate that each important part of the kiln equipment has been inspected or test run. Failure to check each system and starting a kiln under the assumption that all units are ready to go or worse yet, firing-up a kiln in the absence of preestablished schedules, is bound to produce erratic starts and possibly lead to premature shutdowns before the kiln is able to reach operating temperatures. Such complete system checkouts are absolutely essential when the kiln has been shut down for a long period of time (weeks/months) or when major maintenance overhauls have been made.

Each kiln system is unique and different in its design and thus has its own multitude of items that should be included in the aforementioned checklist. Hence, not much benefit could be derived from compiling a list of items in this book that must be checked when those items would not apply to all kilns.

The general format of such a checklist, shown in Table 21.1, could be used by the reader to develop a list that would include all the applicable kiln equipment under his control.

b) Drying of Refractory Lining. Special attention must be given to the drying time when new monolithic (cast) refractory linings have been installed. These linings have to be slowly and completely dried before the temperature of the lining can be appreciably raised. But, this attention toward proper drying should not be limited to monolithic linings alone but should include all refractories that contain appreciable amounts of moisture. Such moisture might have been acquired during storage or during installation, when the wet-laying method (bricks installed with mortar) was used. Finally, there is the ever-present possibility that the lining might have acquired moisture due to condensation in the kiln interior or for accidental reasons during the shutdown (e.g., someone forgot

TABLE 21.1

KILN START CHECKLIST
(proposed format)

Code	Responsible							Location	Item	Checked	By
	A	B	C	D	E	F	G				
1		x						Cooler:	Cooler grates and hammermill clear		
2		x							U.G. compartments, drag chain clear		
3		x							All cooler doors closed and sealed		
4			x						Lubrication completed and lubrication system on		
5		x							All cooler instrumentation functioning		
6	x								Cooler drive inspected and test run		
7	x								Cooler fans inspected and test run		

REMARKS

to shut off the water to the back-end water spray or oxygen-analyzer tube).

The drying of the lining is a separate operation and precedes the normal start-up schedule. Drying can take as long as 5 days on relined preheater and precalciner cyclones or can be as short as 4 h in cases where only a short section of lining has been replaced. This drying is usually accomplished by means of an auxiliary gas or oil burner with a relatively small fire. Burning of a scrap wood pile in the burning zone is still being practiced in many plants to achieve the same results. During the entire designated drying period the kiln back-end temperature is usually held below 400 F (204 C) and firing is done predominantly with the kiln's natural draft only (I.D. fan down). *Note of caution:* Here too, the rules for complete combustion (the right amount of air and sufficient heat for ignition of the fuel) apply at all times. Many operators test for proper kiln draft by holding a rag in front of the burner-hood viewing ports prior to lighting the drying fire. This is a crude but very effective method for preventing lighting up with excessive or insufficient kiln draft.

c) Expansion and Volume Changes. Each type of refractory exhibits different linear-expansion and volume changes during the time it is brought to operating temperature. Such differences can even exist within individual refractory classes such as basic bricks. This is the reason why there are such widespread differences of opinion on the question of how fast or how slow a kiln should be brought to operating temperature. What complicates matters further is the fact that most kiln linings consist of five or more types of refractory, each undergoing different expansion and volume changes and thus, in theory, requiring their own optimum heating schedule. A slow heating schedule that might be beneficial for the basic lining in the burning zone might, at the same time, be destructive to the lining in the upper part of the kiln. Since basic refractories undergo the most and fireclay the least changes during the heating period, it makes sense to develop a heating schedule that predominantly takes into consideration the properties of the burning-zone lining.

No attempt is made herein to discuss the initial heating of a *new* kiln since such starts demand longer periods for heating the kiln and are governed by the manufacturer's specifications. Kiln and refractory manufacturers usually supervise and specify heating schedules for the initial start-up of a new kiln. For these reasons, subsequent discussions will center on the topic of starting a kiln after refractory linings have been replaced.

Several refractory manufacturers recommend that a basic lining should be rapidly heated in the first hours at a rate of 100–150 C/h (212–300 F/h) until 350 C (662 F) is reached followed by a holding period of 2–4 h at this temperature. This is to prevent the possible formation of the destructive hydration products $Ca(OH)_2$ and $Mg(OH)_2$. After this holding period the temperature is usually slowly raised at a rate of 30–60 C/h (50–100 F/h) until operating temperatures are reached. There is widespread agreement that the temperature should not be raised faster than a maximum 100 C/h (180 F/h) after the aforementioned holding period to prevent thermal spalling.

During a start-up, an operator must not only concern himself with the proper expansion of the lining but must also keep close attention to the expansion of the kiln shell and tires since both of these are located at a different distance from the heat source and thus exhibit different expansion behaviors. The important procedure is to allow both to expand at the proper rate such that there can be no constriction of the kiln shell by the slower expanding tires. Extra time spent in closely monitoring tire clearance during a start-up period can prevent the all too frequently observed damages of kiln shells near the tires. Plants, where tire clearance during normal operation is already close to minimal levels should, as a precautionary measure, adopt a somewhat slower heating schedule to prevent any possible damage to the kiln shell.

Periodic turning of the kiln during the heating period is an absolute necessity to prevent an uneven heating of the kiln's circumference, i.e., prevent "one-sided" expansion of the lining and kiln shell. Linings with a low coefficient of thermal expansion (fireclay and insulating bricks) pose the distinct possibility that they could become loose during the first half of the start-up period hence too frequent turning in the first few hours should be avoided.

d) Timing of the Feed Introduction. Feed should not be given to the kiln until the back-end temperature is at a level that will allow the feed to undergo the normal reactions such as evaporation (wet kiln), preheating, and calcining. With very few exceptions, kiln-feed introduction begins when this temperature is between 17–40 C (30–75 F) below the target normal-operating temperature. The feed ratio (feed rate vs. kiln speed) is generally set at 80–90% of desired full production rate to secure a smooth and steady kiln start. Full production rate is normally not attempted until after the kiln has stabilized at full speed at the initial lower feed-ratio setting.

It is not necessary to first reach the operating target in back-end temperature before feed is given to the kiln because kiln speed and feed rates are deliberately set lower during the final stages of a kiln start.

e) Timing of the Burning-Zone Temperature. Ideally, one would like to have the burning zone reach clinkering temperature at the precise moment when the fresh feed arrives in this zone. This is to secure immediate formation of coating and prevent unburned "raw" feed from entering the cooler. Unfortunately these ideal conditions are frequently not met because such timing takes skill and is often a hit-and-miss situation. Between the time the feed is given to the kiln and the time it arrives in the burning zone (considered residence time), an operator should make every effort and closely monitor the kiln to try to achieve this ideal condition for this is the key to any smooth kiln start and produces stable kiln operations thereafter.

Figs. 21.1–21.3 are examples of kiln-start schedules that are typical for the types of kilns indicated. The reader can use these sample schedules, by making the necessary modifications, to develop his own schedule suitable for the kiln under his control. Kiln-start schedules should all have guide-lines for both the burning-zone and back-end temperatures. A review of past kiln-start performance charts can also shed some light on the question of what the fuel rates at any given time are and when the main fire should be lit in the kiln. Such information, if available, should definitively be part of any kiln-start schedule.

21.2 KILN STOPS AND SHUTDOWNS

There are many reasons why kilns periodically have to be shut down but unquestionably the most frequent reasons are:

a) refractory failure,
b) feed-system failure or feed shortages, and
c) clinker cooler-component failures.

These three conditons, in most cases, demand a decision from the operator for an immediate unscheduled shutdown of the kiln system. There is often not enough time available in such situations to weigh or discuss the merits of a shutdown because a few minutes delay could lead to major equipment damage. Clearly, an operator should be fully trained in what to do when such situations occur.

Mistakes have been made by inexperienced operators in such stress situations that have led direclty to costly damages to the equipment. On

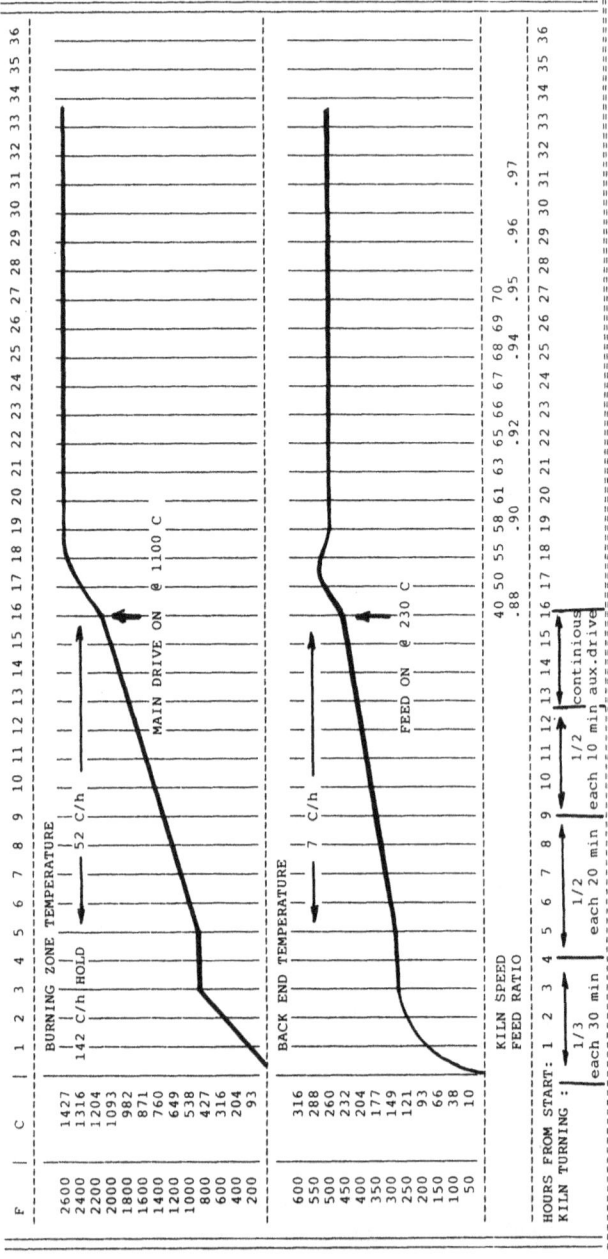

Fig. 21.1 Kiln-start schedule (example). Type of kiln: Wet process < 3.6 m (12 ft) diameter; Targets: Back-end temperature: 260 C (500 F), kiln speed: 70 rph, feed ratio: 0.97

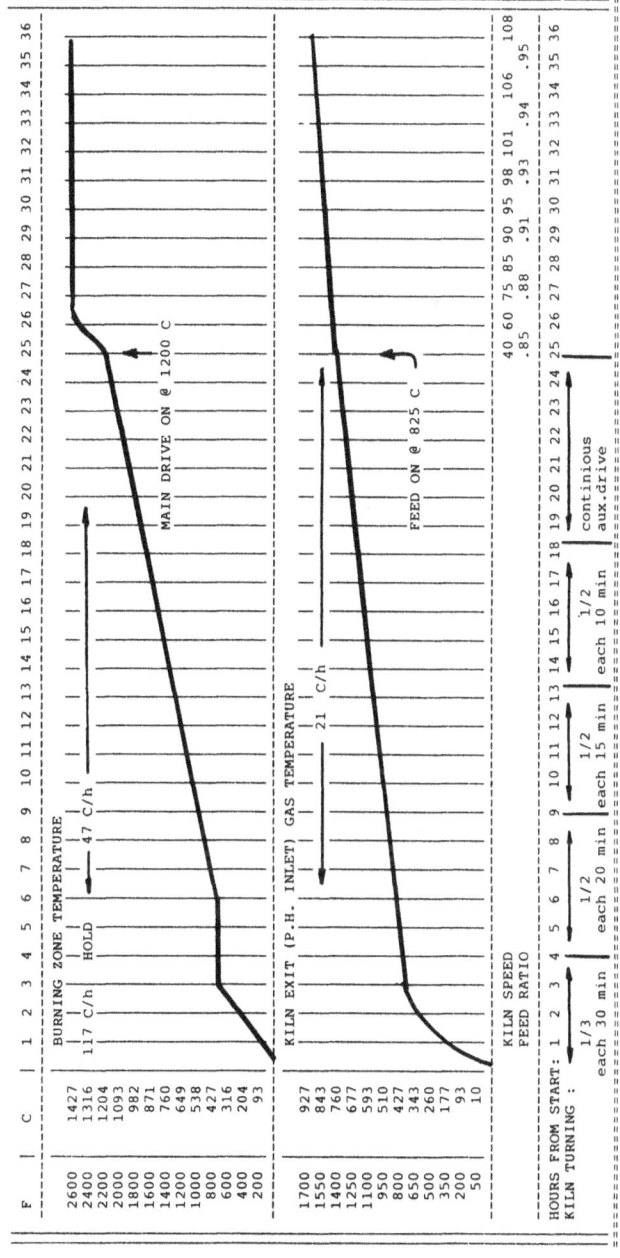

Fig. 21.2 Kiln-start schedule (example). Type of kiln: 4 stage preheater 4.4 m (14.5 ft) diameter; Targets: Kiln exit-gas temperature: 900 C (1650 F), kiln speed: 108 rph, feed ratio: 0.95

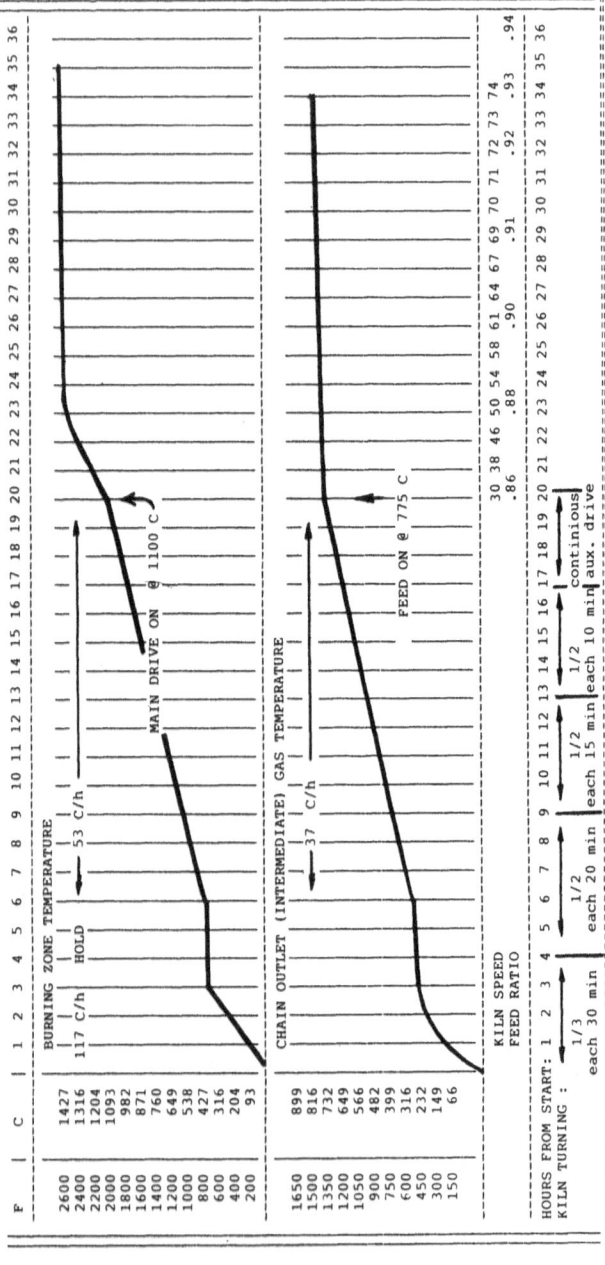

Fig. 21.3 Kiln-start schedule (example). Type of kiln: Dry process > 4.5 m (15 ft) diameter; Targets: Chain outlet temperature: 800 C (1480 F), kiln speed: 74 rph, feed ratio: 0.94

the other hand, it must also be mentioned that many operators have saved their company production delays and costly equipment repairs by keeping their composure and using common sense on many occasions when the kiln seemed to get out of control.

Just as during a kiln start, there are several extraordinary conditions an operator has to be concerned with when shutting a kiln down. These are:

a) Proper cooling schedules to allow the kiln and refractory lining to contract in a predetermined manner.

b) Proper kiln-turning schedules to allow the cooling to proceed evenly on the circumference of the kiln.

c) Emptying of the fuel (especially when firing coal), the feed, and dust-collecting systems prior to the complete shutdown of the kiln.

d) Emptying of feed from the kiln, situations permitting, and in some instances shooting the coating before appreciable cooling has taken place.

e) Protection of the burner pipe, kiln nose castings, and kiln-hood area from the possible excessively high temperatures.

Here too, written standard operating procedures to cover these items are of great help to the operator because kiln shutdowns, hopefully, are not frequent occurrences. As with any extraordinary kiln situation, operators tend to quickly forget the sequence of procedures they should follow and must therefore have some written material at their disposal to periodically review these procedures.

At this point, it is important to examine these items in detail.

a) Thermal contraction of the kiln. A 160 x 4.7 m (525 x 15.5 ft) rotary kiln contracts approximately 0.4 m (16 in.) in its length and approximately 12 mm (0.5 in.) in its circumference during cooling. Likewise, a ring of refractories also contracts during cooling although at a different rate than steel. Too fast a cooling rate can induce thermal stresses on both the lining and the kiln components, hence, the rate and manner in which a kiln is cooled becomes of paramount importance.

Short clinker inventories or tight production schedules often force operators to limit kiln shutdowns to as few hours as possible. Many operators think that so-called forced (fast) cooling of the kiln with the help of the I.D. fan is a simple means to reduce the time requirements for allowing workers to enter the kiln. In doing so they tend to forget that such rapid cooling could produce costly thermal damage to the kiln system.

Although no standard procedures will apply to all types of kilns, the author has found that adherence to the following guidelines can minimize the risk of thermal shock to the equipment.

(1) Kiln-cooling schedules should provide for a gradual cooling of the burning-zone refractory temperature at a rate of not more than 100 C (180 F) per hour.

(2) Smaller kilns (< 4 m diameter) should be cooled at the above stated 100 C/h rate whereas larger-diameter kilns should not be cooled at a rate higher than 75 C/h (135 F/h).

(3) The I.D. fan should be shut down immediately as soon as the fire is taken off (fuel cut-off) and should not be used for forced cooling of the kiln in the first 6 h after the kiln shutdown. This is an absolute must on dry- and wet-process kilns equipped with internal heat exchangers such as chains.

(4) When the fire is taken out, I.D. fan inlet dampers (kiln-draft control dampers) should be closed as much as permissible during the first 3 h to prevent excessively high, natural kiln draft (natural draft occurs without the I.D. fan running).

These guidelines will immediately raise the question of the possibility that positive hood pressures might occur that are also destructive to the kiln equipment. To counteract such conditions, the operator must immediately reduce the total amount of air given to the clinker cooler compartments as soon as the I.D. fan is shut down and/or the kiln-draft damper is being closed.

Note of caution: Make sure the kiln draft is not reversed. In other words, regulate the dampers such that the kiln gases never are drawn from the burning zone into the cooler and the cooler stack.

b) Kiln-turning schedule. To secure uniform cooling, the kiln has to be turned on a regular schedule because the feed bed and the refractory underneath it cools much slower than the refractory wall that is exposed to the kiln gases. An example of such a turning schedule is shown in Chapter 3. Turning schedules should not be compromised in the first 2 h of a shutdown. In other words, when the time comes to turn the kiln, workers should get clear of the trunnions, preheater, and the cooler interior to allow the operator to proceed with the turn.

Note: Make sure they are indeed clear before activating the kiln drive.

On any prolonged kiln shutdown an operator must also make provisions for evacuating the material from the cooler that is being discharged from the kiln with every turn. One doesn't want to be confronted with a situation where the cooler drive can not be started because of a material overload in the cooler.

c) Emptying of the coal, feed, and dust-collecting systems. Time and conditions permitting, all these systems should be emptied out before the fire is taken out of the kiln. This requires a preplanned sequence of events that have to be factored into a kiln shutdown schedule to account for the activities covering the period of 1–2 h before the fire is taken out. Emergency conditions excluded, a gradual, sequenced reduction of the kiln control variables prior to "fire-off" is always preferable to a sudden large change in these controls.

d) Emptying of the feed from the kiln interior. It takes special skill and experience to "burn out" a kiln before the fire is taken out and the kiln drive stopped. Emptying a kiln of feed before the actual shutdown unquestionably makes it easier to work in the kiln interior especially when refractory repairs have to be done in the upper part of the kiln. It must however be emphasized that only experienced operators should be allowed to "burn" a kiln out, because any miscalculation on the part of the operator could lead directly to overheated conditions, either at the back end or the cooler. So-called chain and coal-mill fires have occurred as a direct result of inexperienced operators attempting to burn out a kiln.

The best one can hope for is to burn out all the feed behind the burning zone. Emptying the burning zone itself while the main fire is still lit is far too risky and should never be attempted.

During the entire period of a burnout the operator must adhere to a firm set of rules and must abort his attempts (i.e., immediately shut the system down) whenever any one of these rules are not met. The rules are:

(1) Never allow the oxygen in the kiln exit gases to exceed 1%.
(2) Never allow the back-end temperature (I.G.T. in kilns with chains) to exceed a predetermined maximum limit.
(3) During the entire period of a burnout, the I.D. fan, fuel, kiln speed, and cooler air flows must be progressively reduced to maintain a continuous, gradual decline in back-end temperatures.

e) Protecting the burner pipe. The hood and the kiln discharge area remain at elevated temperatures for several hours after the fire has been taken out of the kiln. Some kilns are equipped with movable burner pipes to allow for partial or full retraction of the burner. Others have to rely on the primary air fan to provide the necessary cooling of the burner during the first 5 h after a kiln shutdown. In these cases, it is advisable to leave the primary-air fan running at reduced speed until such time as the kiln interior has cooled to full blackness.

21.3   KILN CYCLES

One of the most common kiln upsets is the condition in which the back-end and burning-zone temperatures deviate periodically and by a large degree from the optimum range, forcing the operator to execute a kiln slowdown in regular intervals. The time span of each cycle usually is identical to the time it takes for the feed bed to travel from the preheat to the burning zone. Cycling kiln conditions can extend from four to six cycles and, in extreme situations, last as long as several days.

Wet- and dry-process kilns with chain systems are especially subjective to these cycles. Preheater and precalciner kilns very seldom undergo cycles because in these kilns calcining is being done predominantly outside the rotary kiln proper and can be much more easily controlled. Kiln cycles are the result of irregular calcining conditions within the kiln system. Calcining conditions are changed when:

a) feed hold-back occurs in the chain system
b) the feed bed becomes fluidized in the calcining zone
c) the kiln is being fed at irregular feed and fuel rates, and
d) most important of all, when the operator, under any one of these conditions, is not capable of controlling the back-end temperature within the desired operating range.

Fig. 21.4 is a classic example of a cycling kiln and will be used in the ensuing discussions to explain what takes place during a cycle and how to work oneself out of one. In Fig. 21.4 the key variables and controls that govern the kiln cycles can be recognized. They are: BZT (burning-zone temperature), and BET (back-end temperature), both of which are primarily controlled by the kiln speed, I.D. fan, and the fuel rate. The prevailing conditions in this figure are also detailed in Table 21.2.

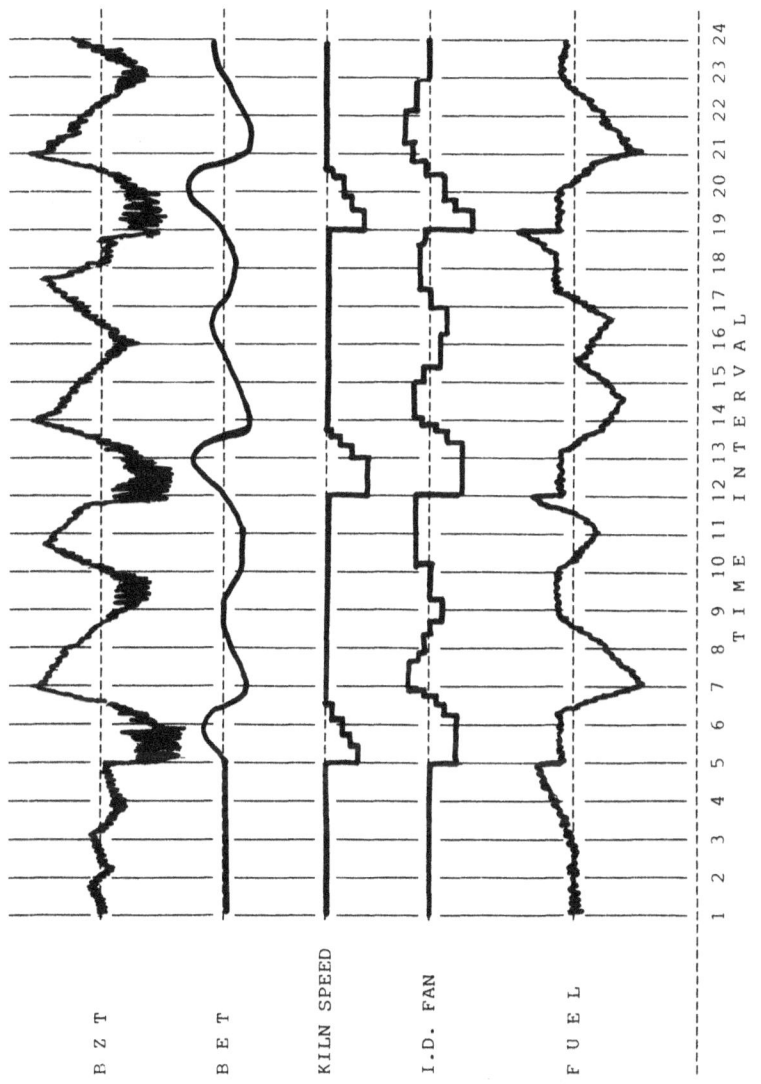

Fig. 21.4 Typical behavior of cycling kilns.

TABLE 21.2

TYPICAL CONDITIONS DURING KILN CYCLES
(as shown in Fig. 21.3)

Time interval	BZT	BET	Kiln speed Status	Kiln speed Action	I.D. fan Status	I.D. fan Action	Fuel rate Status	Fuel rate Action
1	OK	OK	normal		normal		normal	
2	OK	OK	normal		normal		normal	
3	OK	OK	normal		normal		normal	
4	down trend	OK	normal		normal		high	increase
5	too low	OK	low	decrease	low		high	decrease
6	up trend	too high	normal	increase	high	decrease	normal	
7	too high	too low	normal		high	increase	low	decrease
8	down trend	low	normal		low	decrease	low	increase
9	low	OK	normal		normal		high	
10	OK	low	normal		high	decrease	normal	
11	high	low	normal		high		low	increase
12	too low	OK	low	decrease	low	decrease	high	decrease
13	up trend	too high	normal	increase	high	increase	normal	
14	too high	low	normal		high	decrease	low	decrease
15	down trend	too low	normal		high	decrease	low	increase

The time intervals in this figure could be expressed in increments of 10, 20, or 30 min depending on the length and diameter of a given kiln but doesn't alter the indicated typical cycling behavior of these kiln variables.

Cycles are characterized by a heavy and sudden onrush of feed into the burning zone (time intervals: 5, 12, 19) forcing the operator to execute a kiln slowdown. After a given period of operating the kiln at this lower speed, wherein the visibility in the burning zone is usually extremely poor, the kiln suddenly and rapidly heats up. Once visibility improves, the operator is normally confronted with a condition in which the load in the burning zone is very light and dangerously overheated (time intervals 7, 14, 21). This in turn compels the operator to make drastic adjustments in fuel rate and I.D. fan speed. Under certain conditions it might be necessary to increase the kiln speed above the normal target to prevent balling up of the clinker bed in the burning zone.

Such changes are dictated by the prevailing dangerous conditions in which an operator usually has no other choice than to go to a slowdown procedure to prevent raw feed from entering and damaging the clinker cooler. Likewise, afterwards the drastic fuel cuts at intervals of 7, 14, and 21 are necessary to guard against possible overheated conditions in the burning zone that could melt the coating and damage the refractory.

After a kiln has gone through several of these repeat cycles, an operator could easily become thoroughly disgusted with the kiln or become resigned in the belief that such instability is inevitable and that there is not much that could be done to eliminate the problem. There are isolated cases where this might hold true, such as when the kiln-feed properties, kiln speeds, and feed rates are not compatible to the design of the kiln and the chain section. However, such cases are more the exception than the rule. The majority of these cycles are caused by control actions that are focused on the immediate problem without due regard for the aftereffects these actions might have on kiln stability.

In the cited example, the kiln slowdown at time interval 12 is clearly the result of the low back-end temperature at point 7 combined with the high burning-zone temperature (and consequent low fuel rate) at point 11.

The solutions to cycling kilns is to first study the recording charts to determine if there is a clear trend in burning-zone temperature behavior, i.e., determination should be made as to what future time interval a drastic change can be expected. The tip-off in this respect is the back-end temperature at a given previous time interval. Once this time lag and trend is known, kiln control must become a matter of anticipation, i.e., making control adjustments before the key variables (BZT and BET) have a chance to undergo the cycling behaviors.

Assume that the present kiln operation is represented by point 10 in our example. Assume also that the kiln has gone through several cycles and a clear trend has been established. From past trends it should become apparent that, despite prevailing normal burning-zone temperatures and normal settings for I.D. fan and fuel rate, a short period of higher BZ temperatures followed by a "push" can be expected shortly because of the low BET at point 7. Equally important is the observation that the BET is trending lower at this point 10. This is then the opportune time for a change in control strategy to break the cycle.

The author has successfully broken many of these repeat kiln cycles by employing somewhat contrarian procedures when similar conditions to those shown in time interval 10 prevailed. It is important that these preventive adjustments are made BEFORE there is a visible sign of a drastic change in burning-zone conditions. These control adjustments are (at point 10):

1. REDUCE THE FEED RATIO TO REDUCE THE OVERALL KILN OUTPUT UNTIL THE KILN IS AGAIN STABILIZED (i.e., if during previous cycles the feed ratio was set at 0.92, set the ratio to 0.86).
2. REDUCE KILN SPEED TO HALF AND EXTEND THIS SLOWDOWN PERIOD TO DOUBLE THE TIME OF PREVIOUS SLOWDOWNS.
3. GREATLY REDUCE THE I.D. FAN SPEED AND MAINTAIN THE PREVAILING LOW FUEL RATES. (Note that one can expect the BET to automatically go up because of the lower kiln speed.) Keep a close eye on the oxygen and try to maintain a slightly hotter-than-normal BZT for the next 7 intervals.
4. FOR THE NEXT SEVEN TIME INTERVALS, MAKE EVERY EFFORT TO STABILIZE THE BACK-END TEMPERATURE AND MAKE SURE IT DOESN'T EXCEED THE TARGET LEVEL (although difficult to attain, it is advantageous to try to maintain a BET slightly lower than normal).

All four steps are taken simultaneously. These adjustments are contrary to normal control procedures but necessary for a successful reestablishment of stable kiln operations. The feed ratio should not be raised to the previous levels until the kiln has operated at least 8 h without a kiln

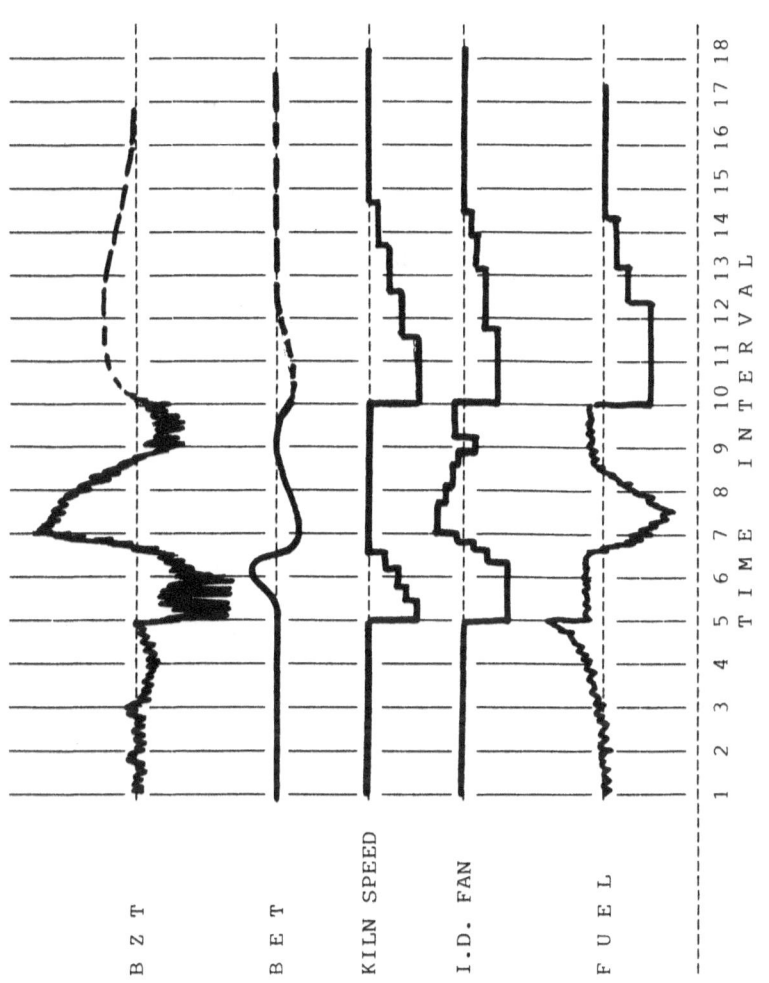

Fig. 21.5 Control actions to combat cycling kilns.

slowdown. The aforementioned principles for breaking a kiln cycle are shown in Fig. 21.5. There is no guarantee that these procedures will work all the time but they are certainly preferable to "fighting" a cycling kiln for hours or even days.

It should be remembered as mentioned earlier, that kiln cycles can be caused by faulty design of chain systems both in wet- and dry-process kilns. Experience has shown that certain systems tend to retain the feed in the chains (feed hold-back) when the kiln is operated at slow speeds. Once the kiln is raised to normal levels, the system will release the accumulated load which in turn will arrive $1-1^1/_2$ h later in the burning zone in form of a "push." When there are definite indications that these are the reasons for cycling kiln conditions it is advisable to keep *very low* kiln speed periods to a minimum, i.e., get back to say 45 rph as quickly as possible

22.

The 27 Basic Kiln Conditions

In the previous discussions on burning-zone and back-end temperature control, reference was repeatedly made to the necessity of not operating a kiln by giving attention to one single kiln variable while allowing the other variables to arbitrarily move about the recording charts in a random fashion.

The ultimate aim of efficient, stable kiln operations can only be attained when an operator continuously strives to achieve steady-state conditons at both ends of the kiln, i.e., maintaining both the burning-zone and the back-end temperatures within a narrow range. Making a control adjustment to alleviate a problem at one end of the kiln and neglecting the effect this adjustment might have at the other end will almost certainly lead to upset and unstable conditons later.

This important principle of kiln operation can be emphasized using two examples. In these examples, "actual" refers to the present kiln situation and "target" represents the optimum condition, i.e., what the temperatures should be.

Example 1.	*Actual*		*Target*	
	F	C	F	C
Burning-zone temperature	2560	1403	2700	1482
Back-end temperature	1550	843	1500	816
Oxygen (%)		2.5		1.0–1.5

Before reading on, it would be helpful to ask oneself the question, "What would I do in such a situation?" allowing oneself the time to think it over.

Real kiln conditions are usually not as simple as those depicted in these examples, but taking herein a simplified approach is helpful in illustrating the important aspects discussed in this chapter. Assuming that there is no major imminent upset in the making one should formulate a course of action and then read on.

Clearly what prevails in this example is a *low* burning zone, a *high* back-end temperature, and moderately *high* oxygen content. If your answer was: "Increase the fuel," you clearly forgot to take the back end into consideration. Increasing the fuel in such a situation would correct the burning-zone temperature but would also drive the back-end temperature higher which, in this example, is definitely not desirable for the sake of kiln stability. The answer should be: "Decrease I.D. fan." This single adjustment would correct three deviations from target at the same time. Lowering the I.D. fan speed draws less heat to the rear of the kiln and thus increases the burning-zone temperature, decreases the back end, and also decreases the oxygen content. The answer: "Increase fuel and decrease I.D. fan," is also acceptable in this example provided that these actions don't drive the oxygen too low and start to show combustibles.

Example 2 is presented without explanation and can be used by the reader to test his grasp of the fundamental principles discussed above.

Example 2.	Actual		Target	
	F	C	F	C
Burning-zone temperature	2740	1504	2700	1482
Back-end temperature	1470	799	1500	816
Oxygen (%)		0.6		1.0–1.5

The examples shown above have been used to demonstrate to the reader the basic, simplified thinking process that must go through an operator's mind before making an adjustment. It is always necessary to check:

1. the burning-zone temperature,
2. the back-end temperature,
3. the percent oxygen in the kiln exit gases,

and then and only then, make a decision about which of the controls (fuel, I.D. fan, and/or kiln speed) needs adjustment. This principle of control applies to any kiln, whether it is a wet, dry or preheater kiln. It even applies to precalciner kilns, but since they have two firing locations, they require application of this principle to both parts separately, namely the rotary kiln and the preheater tower.

Most kiln conditions are not quite as simple as depicted in the above two examples. There are many more complex problems an operator can be confronted with that will test his special skills. A more complex situation is shown in Example 3:

Example 3.	Actual		Target	
	F	C	F	C
Burning-zone temperature	2750	1510	2700	1482
Back-end temperature	1600	871	1500	816
Oxygen (%)		0.7	1.0–1.5	

It is readily apparent that the back-end and burning-zone temperatures are too high, and the percentage of oxygen is below its permissible minimum. The speed of the I.D. fan cannot be reduced as this would result in an even lower oxygen content and most likely would lead to incomplete combustion of the fuel. However, one favorable factor is that the burning-zone temperature is higher than desired. This is important, because a well-heated burning zone in this instance is an absolute requirement in order for corrective measures to be carried out. The first step is to reduce the fuel rate slightly which will increase the oxygen. The burning zone will now cool down if no further steps are taken. Therefore, as soon as the oxygen has increased, reduce the I.D. fan speed in such a way that the oxygen will again return to the previous 0.7%. The back-end temperature will now decrease and theoretically the same burning-zone temperature will be maintained because with lower I.D. fan speed less heat is transferred to the rear.

In order to reduce the back-end temperature 100 degrees, the above described procedure has to be repeated several times in small steps, making sure to allow for resting periods in order that the reactions can be properly observed and overcontrolling avoided. The burning zone must remain well heated throughout the whole procedure. Whenever a cooling down of the burning zone takes place, the procedure has to be stopped until the burning zone has warmed up again.

The example is based on the assumption that the kiln has been operating at full speed for several hours before any attempt is made to lower the back-end temperature. Should the kiln speed be lower than full operating speed, this procedure will not be suitable because any kiln speed increase would automatically lower the back-end temperature. This example has been given for the sole purpose of showing how the back-end temperature can be lowered even when the kiln has been operating under fairly stable conditions for many hours or even days. It is a fact that many rotary kilns are operating unnoticed under excessivley high back-end temperatures. Low oxygen contents do not necessarily mean that one should attempt to operate the kiln with high I.d. fan speed and fuel rate.

Now consider another example in which the back-end temperature is 1400 F (760 C), burning-zone temperature is 2800 F (1540 C) and oxygen is 1.5%. In this case again assume that the kiln in question is operating at full speed. The burning-zone temperature and the oxygen both are at the desired level but the back-end temperature is lower than required. First increase the I.D. fan speed slightly. This will cause the back-end temperature and the oxygen to increase. The burning-zone temperature will now decrease if no further steps are taken. As soon as the oxygen indicator shows a higher reading, increase the fuel rate in such a fashion as to allow the oxygen to again return to the previous level of 1.5%. This fuel increase will compensate for the heat removed to the rear of the kiln by the I.D. fan-speed increase.

Here again one has to make sure that the procedure is undertaken in small steps and if any change in the burning-zone condition is observed, the procedure has to be stopped until the burning zone has again returned to normal.

In this chapter, the terms ideal and proper back-end temperature have been used, which raises the question: "What is the correct back-end temperature and how can this correct temperature be determined?" This cannot be readily answered because many factors such as feed rate, kiln speed, feed composition, moisture content of the feed, and others, all have an influence on this back-end temperature. However, certain conditions are known to be the direct result of incorrect back-end temperature. The following are signs that the back-end temperature is too high:

a) Kiln is operating continuously with a high oxygen content in the exit gases.

b) Continuous easy-burning conditions are identified by the condition in which clinker formation starts too far back from the flame. The burning zone is unusually long.

c) Past kiln performances reveal that the kiln has been operating

under stable conditions previously for prolonged times with a
lower back-end temperature, under almost identical conditons of
kiln speed, feed rate, feed composition, etc.

d) Kiln is operating at poor fuel efficiency, more fuel being used
than would normally be requried to burn the same amount of a
given feed composition.

e) There is lower than normal feed moisture of the sample taken
from the downstream end of the chains in a wet-process kiln.

A back-end temperature that is too low might be indicated by the
following:

a) Kiln operates for an extended period of time at low oxygen
content in the exit gases.

b) Difficult burning conditions for prolonged time caused by partially calcined feed entering the burning zone.

c) Higher than normal moisture content of the feed sample taken
from the downstream end of the chains in a wet-process kiln.

It is important that the kiln is in a relatively stable conditon before
attempting an investigation into the back-end temperature conditions.
Studies of past kiln performance by means of kiln logs and recorder charts
will reveal valuable information as to whether a kiln is operating at settings that are very close to previous, stable long-time runs.

The reader is reminded that a close control over the back-end temperature
is essential to maintain stable kiln conditions.   Chances of obtaining
stable kiln conditions are practically nil if the back-end temperature is
allowed to fluctuate freely.

By now the reader must realize that a kiln cannot be expected to operate
indefinitely in a state of equilibrium. Fluctuations do occur, regardless of
whether a kiln is manually or automatically controlled, and it is the
operator's responsibility to see that the proper adjustments are made
whenever conditions in the kiln vary outside the range of acceptability.
Although there are many variables, with a multitude of combinations, it
has been found that there are three key variables that are of prime
importance. Except under emergency or upset conditons, the operator will
find that these three variables can be maintained within reasonable limits
by means of adjustments to one or more of the three basic controls.

22.1 THE THREE BASIC VARIABLES AND CONTROLS

Conditions in the kiln are indicated by:

a) The burning-zone temperature, which is the dominant influence on the quality of the product.
b) The back-end temperature, which is the principal control on operational stability, and
c) The percentage of oxygen in the exit gas, which governs combustion conditions and fuel efficiency.

Because any of these conditions can be within the allowable range, below the minimum allowable value, or above the maximum value, there are 27 possible basic conditions that will be encountered by the kiln operator. These are shown graphically in Fig. 22.1. Note that a case number has been assigned to each condition as an aid to identification.

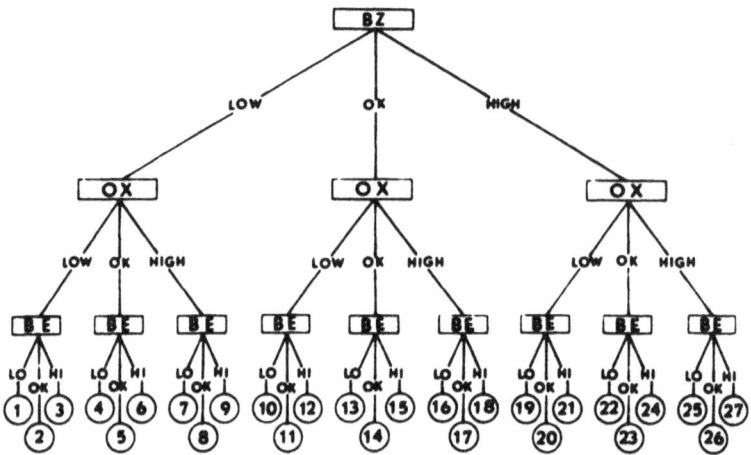

Fig. 22.1 Evaluation of the three basic variables, burning-zone temperature (BZ), percentage of oxygen in the exit gas (OX), and back-end temperature (BE), identifies the case number which can be found in Table 22.1 to assist the operator in controlling his kiln.

Basic Controls. In most conditions, except for the emergencies and upsets listed, control of the three basic variables can be effected by adjustments to the rate of admission of fuel to the burner, changes in the speed of the kiln, and changes in the speed of the I.D. fan.

Application Of Control Procedures. First, it is necessary to establish targets for the three variables: temperature of the burning zone, back-end temperature, and percentage of oxygen in the exit gas. This information will be supplied to the operator by his supervisor, and depends on the composition of the raw feed, type of clinker being burned, and other factors. Similarly, ranges are established through which these variables can safely be permitted to fluctuate, because it is obvious that any variable cannot be held exactly on target at all times. Once these targets and ranges have been established, then Fig. 22.1 and Table 22.1 can be used to aid in determining the proper procedure to correct an out-of-control condition. Abbreviations used in the table and figure are: BE, back-end temperature; BZ, burning-zone temperature, and OX, percentage of oxygen in kiln exit gas. Of course, good judgment is necessary in evaluating the conditions, making sure that no upset or emergency conditions exist.

Example: Assume that in a certain kiln the following values have been established:

Item	Code	Target	Range	Limits
Burning-zone temperature	BZ	2800 F	±50	2750–2850
Back-end temperature	BE	1450 F	±20	1430–1470
Oxygen percentage	OX	1.2%	±0.8	0.4–2.0%

Then BZ is low if below 2750; OK at 2750–2850; high if above 2850. BE is low if below 1430; OK at 1430–1470; high if above 1470; OX is low if below 0.4; OK at 0.4–2.0; high if above 2.0.

At the time under consideration, assume that burning-zone temperature is 2650 (BZ is low), percentage of oxygen is 2.8 (OX is high), and back-end temperature is 1500 (BE is high). In Fig. 22.1 following the line *BZ low, OX high, BE high*, conditions are found in case No. 9. Under case No. 9 in Table 22.1 the correct action is listed.

A similar procedure is followed for correcting any of the 27 listed conditions. The operator should be alert to remedy any out-of-control condition as soon as the limits of tolerance set up for the particular kiln and materials are reached. Corrections should be prompt, but care must be exercised to avoid overcontrolling which could possibly lead the kiln into a cycling condition.

Emergencies and Upsets. The control procedures described in this section are not adequate when the kiln is in an upset condition, or an emergency exists. Special procedures are necessary under such conditions. Briefly, these conditions are:

a) Rapid formation of a ring
b) Loss of coating or ring
c) Red spots on the kiln shell
d) Balling of clinker in the burning zone (sausaging)
e) Dangerously high back-end temperature
f) Combustibles in the flue gas
g) Loss of kiln feed and uneven feed-bed depth
h) Unburned feed in the cooler
i) Upsets in which the kiln speed is either higher or lower than normal.
j) Kiln startup or shutdown periods
k) Any mechanical or electrical equipment failure that interrupts in any way the regular flow of material, gas, fuel, or air.

Whenever any of these conditions exist, it is necessary to apply the corrective measures recommended in special sections of this book. The reader should refer to them and become familiar with them.

The preceding discussions of the 27 basic kiln conditions were highly simplified to show the thought process an operator must follow when controlling a kiln manually. The advent of the inexpensive microcomputer has made it possible to refine and extend this control concept wherein other variables that are also instrumental toward kiln stability can be included. Computers are capable of scanning all these important variables in a split second and, if properly programmed, can make the necessary adjustments to either maintain or to lead the kiln into stable operations.

The author has developed the software for a training program that is based on the 27 basic kiln-control concept but that also takes into account such other factors as the specific heat input requirements, effects of secondary air temperatures, kiln-drive amperage, and time trends in back-end temperatures. Samples of this control concept are shown in Table 22.2 and Table 22.3.

Table 22.2 is of interest because it represents condition 9 that has been extensively discussed earlier (Example 1) with the significant difference that this time the kiln speed is already below normal therefore demanding different considerations.

TABLE 22.1

KILN OPERATING PROCEDURES

Case	Condition	Action to be taken	Reason
1.	BZ low OX low BE low	When BZ is drastically low: a. Reduce kiln speed b. Reduce fuel When BZ is slightly low: c. Increase I.D. fan speed d. Increase fuel rate	To increase burning zone and back-end temperatures To move the oxygen percentage into range To raise the back-end temperature and the oxygen percentage for action (d) To raise burning-zone temperature and to return oxygen percentage into range
2.	BZ low OX low BE OK	a. Reduce kiln speed b. Reduce fuel rate c. Reduce I.D. fan speed	To raise burning-zone temperature To raise oxygen percentage for action (c) To maintain back-end temperature
3.	BZ low OX low BE high	a. Reduce kiln speed b. Reduce fuel rate c. Reduce I.D. fan speed	To raise burning-zone temperature To increase oxygen percentage for action (c) To reduce back-end temperature (but be sure the BE is in a definite downward trend)
4.	BZ low OX OK BE low	When BZ is drastically low: a. Reduce kiln speed When BZ is slightly low: c. Increase I.D. fan speed d. Increase fuel rate	To raise both the back-end and burning-zone temperatures To raise back-end temperature and oxygen percentage for action (d) To raise burning-zone temperature

TABLE 22.1 (cont'd.)

Case	Condition	Action to be taken	Reason
5.	BZ low OX OK BE OK	When BZ is drastically low and oxygen is in lower part of range: a. Reduce kiln speed b. Reduce fuel rate c. Reduce I.D. fan speed d. Increase fuel rate	 To raise burning-zone temperature To raise oxygen percentage for action (c) To maintain back-end temperature To raise burning-zone temperature
6.	BZ low OX OK BE high	When BZ is drastically low: a. Reduce kiln speed b. Reduce fuel rate c. Reduce I.D. fan speed When BZ is slightly low and oxygen in higher part of range: d. Reduce I.D. fan speed	 To raise burning-zone temperature To raise oxygen percentage for action (c) To lower back-end temperature To reduce back-end temperature and raise burning-zone temperature
7.	BZ low OX high BE OK	When BZ is drastically low: a. Reduce kiln speed b. Reduce I.D. fan speed When BZ is slightly low: c. Increase fuel rate	 To raise both back-end and burning-zone temperatures To maintain back-end temperature To raise burning-zone temperature and lower oxygen percentage. Back-end temperature should be rising also; if not, continue increasing fuel rate, and increase I.D. fan speed
8.	BZ low OX high BE OK	When BZ is drastically low: a. Reduce kiln speed b. Reduce I.D. fan speed	 To raise burning-zone temperature To maintain back-end temperature. If oxygen is still available after this move, increase fuel rate also.

TABLE 22.1 (*cont'd.*)

Case	Condition	Action to be taken	Reason
		When BZ is slightly low: c. Increase fuel rate	To raise burning-zone temperature and lower oxygen
9.	BZ low OX high BE high	When BZ is drastically low: a. Reduce kiln speed b. Reduce I.D. fan speed c. Increase fuel rate When BZ is slightly low: d. Reduce I.D. fan speed	To raise burning-zone temperature To lower back-end temperature and oxygen percentage To raise burning-zone temperature and lower percentage of oxygen To raise burning-zone temperature and reduce back-end temperature and oxygen percentage
10.	BZ OK OX low BE low	a. Increase I.D. fan speed b. Increse fuel rate	To raise back-end temperature and increase oxygen percentage for action (b) To maintain burning-zone temperature
11.	BZ OK OX low BE OK	a. Decrease fuel rate slightly	To raise percentage of oxygen
12.	BZ OK OX low BE high	a. Reduce fuel rate b. Reduce I.D. fan speed	To increase percentage of oxygen for action (b) To lower back-end temperature and maintain burning-zone temperature

TABLE 22.1 (cont'd.)

Case	Condition	Action to be taken	Reason
13.	BZ OK OX OK BE low	a. Increase I.D. fan speed b. Increase fuel rate	To raise back-end temperature To maintain burning-zone temperature
14.	BZ OK OX OK BE OK	NONE. However, do not get overconfident, and keep all conditions under close observation	
15.	BZ OK OX OK BE high	When oxygen is in upper part of range: a. Reduce I.D. fan speed When oxygen is in lower part of range: b. Reduce fuel rate c. Reduce I.D. fan speed	To reduce back-end temperature To raise oxygen percentage for action (c) To lower back-end temperature and maintain burning-zone temperature
16.	BZ OK OX high BE low	a. Increase I.D. fan speed b. Increase fuel rate	To raise back-end temperature To maintain burning-zone temperature and reduce percentage of oxygen
17.	BZ OK OX high BE OK	a. Reduce I.D. fan speed slightly	To lower percentage of oxygen
18.	BZ OK OX high BE high	a. Reduce I.D. fan speed b. Reduce fuel rate slightly	To lower back-end temperature and percent oxygen To maintain burning-zone temperature

TABLE 22.1 (*cont'd.*)

Case	Condition	Action to be taken	Reason
19.	BZ high OX low BE low	When BZ is drastically high: a. Increase kiln speed b. Increase I.D. fan speed c. Reduce fuel rate When BZ is slightly high: d. Increase I.D. fan speed	To avoid overheating the burning zone To raise back-end temperature and oxygen percentage To lower burning-zone temperature and raise both oxygen percentage and back-end temperature.
20.	BZ high OX low BE OK	When BZ is drastically high: a. Increase kiln speed b. Decrease fuel rate c. Increase I.D. fan speed When BZ is slightly high: d. Reduce fuel rate	To avoid overheating To lower burning-zone temperature To increase oxygen percentage and maintain back-end temperature To lower burning-zone temperature and raise oxygen percentage
21.	BZ high OX low BE high	When BZ is drastically high: a. Increase kiln speed b. Reduce fuel rate When BZ is slightly high: c. Decrease fuel rate	To lower back-end temperature and avoid overheating To lower burning-zone and back-end temperatures and increase oxygen percentage To lower burning-zone and back-end temperatures and increase oxygen percentage

TABLE 22.1 *(cont'd.)*

Case	Condition	Action to be taken	Reason
22.	BZ high OX OK BE low	When BZ is drastically high: a. Increase kiln speed b. Increase I.D. fan speed c. Reduce fuel rate When BZ is slightly high: d. Increase I.D. fan speed	To avoid overheating To raise back-end temperature To lower burning-zone temperature To raise back-end temperature and lower the burning-zone temperature*
23.	BZ high OX OK BE OK	When BZ is drastically high: a. Increase kiln speed b. Decrease fuel rate c. Increase I.D. fan speed When BZ is slightly high: d. Reduce fuel rate	To avoid overheating To lower burning-zone temperature To maintain back-end temperature To lower burning-zone temperature*
24.	BZ high OX OK BE high	When BZ is drastically high: a. Increase kiln speed b. Decrease fuel rate When BZ is slightly high: c. Decrease fuel rate	To avoid overheating and lower back-end temperature To lower burning-zone temperature To lower both burning-zone and back-end temperature*

TABLE 22.1 (*cont'd.*)

Case	Condition	Action to be taken	Reason
25.	BZ high OX high BE low	When BZ is drastically high: a. Increase kiln speed b. Increase I.D. fan speed When BZ is slightly high: c. Increase I.D. fan speed	To avoid overheating To increase back-end temperature and lower burning-zone temperature To raise back-end temperature and lower burning-zone temperature*
26.	BZ high OX high BE OK	When BZ is drastically high: a. Increase kiln speed b. Increase I.D. fan speed c. Decrease fuel rate When BZ is slightly high: d. Reduce fuel rate	To avoid overheating To maintain back-end temperature To reduce burning-zone temperature To lower burning-zone temperature*
27.	BZ high OX high BE high	When BZ is drastically high: a. Increase kiln speed b. Decrease fuel rate When BZ is slightly high: c. Decrease fuel rate d. Decrease I.D. fan speed	To lower burning-zone and back-end temperatures To lower burning-zone temperature To lower the back-end temperature*

* If the percentage of oxygen increases during this adjustment, disregard it until the temperatures are brought under control.

TABLE 22.2

OPERATING CONDITION 3A

		Target setpoint	Actual	Deviation	Operating range
Burning zone	(F)	2700	2660	-40	2670–2730
Back-end temperature	(F)	1500	1525	25	1480–1520
Oxygen	(%)		0.3		0.7–2.0
Combustibles	(%)	0	0.1	0.1	0
Back-end temperature	1.5 h ago	1500	1475	-25	
Fuel rate	(lb/h)	23316	25470	2154	± 5%
Heat input	(Btu/lb clinker)	2063	2205	142	N/A
Fuel heat value	(Btu/lb)	12600	12320	-280	
I.D. fan speed	(rpm)	525	513	-12	640–660
Kiln drive	(amps)	31	26	-5	29–33
Secondary air	(F)	1600	1640	40	1520–1680
Kiln speed	(rph)	72	72	0	35–80
Feed rate	(lb/h)	225000	224850	-150	N/A
Clinker output	(lb/h)	142405	142310	-95	N/A

TABLE 22.2 (cont'd.)

Kiln status	Condition 3A	
DANGER—COMBUSTIBLES	BZ low	BE high
Overfueling kiln	OX low	
I.D. fan is on lower speed		
Kiln cooling down		
BZ temperature in downward trend		
Control action		
Immediately reduce fuel rate by 2000 lb		
Decrease kiln speed by 15 rph		
(when combustibles = 0 and OX > 1.0: Decrease I.D. fan by 3 rpm)		

TABLE 22.3

OPERATING CONDITION 9A

		Target setpoint	Actual	Deviation	Operating range
Burning zone	(F)	2700	2640	-60	2670–2730
Back-end temperature	(F)	1500	1550	50	1480–1520
Oxygen	(%)		2.3		0.7–2.0
Combustibles	(%)	0	0	0	0
Back-end temperature	1.5 h ago	1500	1530	30	
Fuel rate	(lb/h)	23316	23860	544	± 5%
Heat input	(Btu/lb clinker)	2063	2561	498	N/A
Fuel heat value	(Btu/lb)	12600	12320	-280	
I.D. fan speed	(rpm)	525	518	-7	640–660
Kiln drive	(amps)	31	34	3	29–33
Secondary air	(F)	1600	1170	-430	1520–1680
Kiln speed	(rph)	72	61	-11	35–80
Feed rate	(lb/h)	225000	181320	-43680	
Clinker output	(lb/h)	142405	114759	-27646	N/A

TABLE 22.3 *(cont'd.)*

Kiln status	Condition 9A		
Overfueling kiln	BZ low	BE high	
Kiln speed below normal		OX high	
Burning zone heating up			
Secondary air below normal			
BZ temperature is in upward trend			
Control action			
Increase kiln speed	3 rph		
Decrease I.D. fan	2 rpm		

23.

Kiln Emergency Conditions

Even in the best-designed and controlled kilns the operator is sometimes confronted with an emergency situation that calls for immediate and drastic actions. There is usually not enough time available to consult others as to the course of action. Quick and appropriate decisions are crucial and must be decisive for the prevention of possible major damage to the kiln equipment. In such an emergency, an operator must ignore kiln-stability maintenance and concentrate on getting full control over the immediately dangerous condition. Clearly, the most important requirement in such situations is that the operator maintains a cool composure.

The following discussions are brief and in a notation format in order to provide the reader with a quick-access reference. An index has been inserted to make it possible for the reader to readily find the subject matters in this chapter. The subjects discussed herein are only a limited sample of the most common emergency conditions that can be encountered on a rotary kiln and are by no means complete. They can serve as the cornerstone for a more extended list of standard operating procedures that must be separately written and prepared for each type of kiln.

No kiln operator training can be considered complete unless the trainee has a full knowledge of what to do when such emergencies arise. Since most likely there is not enough time available to research the procedures or to ask others what to do, the operator's reaction must come naturally and spontaneously.

Management should review the following operating procedures to make user they are applicable to their own kiln system. The adoption of a procedure must be appropriate to the particular kiln.

INDEX OF EMERGENCY CONDITIONS

EMERGENCY CONDITION 23.1:
RED SPOT ON KILN SHELL

Indicators:
- By visual observation
- Shell scanner: Sharp and rapid shell-temperature increase to a level above 450 C (850 F)
- Visual observation of loose refractory bricks in the clinker bed of the burning zone

---

Possible Effects and Dangers:
Severe warpage and damage to kiln shell.

Recommended Action to Take:
A) *Small red spot, located in upper transition or center of burning zone:*

   Continue normal operation of kiln but:
   1. Start shell-cooling fans in the area of the red spot.
   2. Shorten flame to bring "black" feed over area of red spot in an attempt to form new coating.
   3. Maintain normal burning-zone temperatures.
   4. Change kiln-feed chemistry to obtain easier-burning mix.

B) *Large red spot, located under or near a kiln tire or in areas where usually no coating is formed:*

   Shut kiln down immediately

WARNING: Under no circumstances should a water spray be used on the red spot as this could immediately result in severe kiln-shell damage.

Possible Measures to Prevent Reoccurrence:
- Make sure flame configuration and characteristics are not causing localized coating erosion or continuous, excessive overheating.
- Employ proper refractory installation methods.
- Minimize frequency of kiln shutdowns and upsets.
- Avoid "hard"-burning mixes (i.e., ensure sufficient percent liquid content in mix to promote coating formation).

EMERGENCY CONDITION 23.2: RAW, UNBURNED FEED IN CLINKER COOLER

Indicators:
- Onrush of raw feed into and beyond burning zone
- "Black feed" position advanced more than halfway under the flame in burning zone
- "Black-out" in burning zone
- Red grates in cooler
- Rapid rise in cooler grate and clinker discharge temperatures
- Cooler drag-chain amperage increases rapidly

Possible Effects and Dangers:
- Thermal damage to cooler grates and grate-drive mechanism
- Flame extinguishment in burning zone
- Fire on clinker conveyor belts
- Excessively high temperatures in coal-mill air circuit

WARNING: *Watch for incomplete combustion when visibility in burning zone is severely restricted.*

Action to Take: First and foremost—don't wait until raw feed is in the cooler; act when the first signs of impending problems are visible in the burning zone.
1. Immediately reduce kiln speed to minimum (or turn kiln on auxiliary drive).
2. Reduce fuel and I.D. fan speed in accordance with standard slowdown procedures to protect kiln back end.
3. Reduce cooler-grate drive speed (switch to manual control) to allow material in cooler more time for cooling.
4. Adjust cooler air-flow rates to obtain maximum cooling without the hood pressure going positive.
5. Advise all unauthorized personnel to stay clear of the firing floor, cooler, and coal-mill area.

Preventive Measures for Reoccurence:
- Accelerate frequency of visual observation of burning zone for early detection of impending cooler upsets.
- Evaluate kiln output rates vs. cooler capabilities and kiln-operating stability.

EMERGENCY CONDITION 23.3:
LARGE RING BROKEN LOOSE IN KILN

Indicators:
- Visual observation of large junks in burning zone
- Sudden drop in kiln back-end draft
- Large drop in oxygen content of kiln exit gases
- Hood pressure tending toward positive side
- Sudden change in kiln-drive amperage

Possible Effects and Dangers:
- Overloading cooler with unburned feed
- Onrush of excessive amounts of feed into the burning zone
- Damage to cooler drive and grates
- Large pieces jamming cooler hammer mill
- Red-hot clinker leaving cooler

Action to Take:
   When amount of feed and ring fragments in burning zone is extremely large:
1. Immediately reduce kiln speed to minimum.
2. Reduce fuel and I.D. fan speed to keep back-end temperature under control.
3. Switch cooler grate control to manual and reduce grate speed.
4. Adjust cooler air flows to maximum flow possible; without that the hood pressure goes positive.
5. Have personnel on standby to watch the cooler and the hammermill for possible overloading, overheating, and jamming.

Possible Preventive Measures for Reoccurence:
- Laboratory to reevaluate chemistry of kiln feed (including dust-return rates) for possible elimination of ring formation. If no solution in this area possible then:
- Initiate regular schedule to remove rings and heavy buildup by means of special devices designed for this purpose.
- Initiate regular procedures to displace the burning-zone location on a daily basis (i.e., reposition burner every morning).

EMERGENCY CONDITION 23.4:
BURNING ZONE DANGEROUSLY HOT

Indicators:
- Clinker balling ("sausage") in burning zone
- Coating dripping off the wall
- Sliding molten clinker bed in burning zone
- Burning-zone temperature recording too high
- Cooler undergrate pressures too high
- Yellow/white burning zone

Possible Effects and Dangers:
- Loss of coating and thermal damage to refractory
- Red spots on the kiln shell
- Thermal damage to cooler and kiln-hood components

Possible Actions:
1. Reduce fuel rate to minimum until sausaging stops.
2. Increase kiln speed approximately 5–10 rph until sausage is broken
3. Provide maximum possible air in cooler (without hood pressure going positive).
4. Reduce primary air flow.
   THEN, AS SOON AS THE PRIMARY OBJECTIVE OF BREAKING THE AGGLOMERATION IS ACCOMPLISHED:
5. Reduce the kiln and I.D. speed AND increase fuel rate to restore normal operating conditions.

Preventive Measures:
- If "sausaging" is frequent and the result of easy-burning mix, have laboratory evaluate possibility of providing a mix with less percent liquid content.
- Make more frequent, vigilant observation of burning-zone conditions.
- Evaluate flame position and shape to determine if thinner, longer flame is feasible.

EMERGENCY CONDITION 23.5: SUDDEN, SHARP RISE IN BACK-END TEMPERATURE

Possible Reasons:
- Feed shortage
- Combustibles in exit gas
- I.d. fan speed too high
- Kiln speed too low
- Chain "fire"

Possible Effects and Dangers:
- Chain fire on wet and dry kilns
- Thermal damage to back end, dust collector, and preheater tower equipment
- Delayed ignition of fuel in back end of the kiln

Possible Actions:
1. Immediately de-energize electrostatic precipitator.
2. Immediately reduce fuel rate and I.D. fan speed to obtain less than 0.3% oxygen in exit gas.

WARNING: *Do not cut off fuel rate completely as this could trigger an explosion.*

3. Increase kiln speed and feed rate.
4. Warn personnel to stay clear of kiln back end.
5. Do not open any doors in kiln back end.
   THEN, AS SOON AS THE PRIMARY OBJECTIVE OF BRINGING THE BACK-END TEMPERATURE UNDER CONTROL IS ACCOMPLISHED:
6. Return kiln control variables to normal to restore normal operating conditions.
7. Check out back end to determine if thermal damage has occurred.

Preventive Measures:
- Do not operate kiln without feed for more than 10 min.
- Provide alarms and properly maintain kiln instrumentation to obtain warnings before the back-end temperature gets out of hand.
- Maintain close vigilance over combustion, back end, and feed-flow conditions during kiln starts, shutdowns, and upsets.

EMERGENCY CONDITION 23.6:
BLACK SMOKE EMISSION FROM KILN STACK

Indicators:
- Combustibles in exit gases
- Oxygen in exit gas too low
- Flame extinguished for poor ignition conditions
- Burning-temperature too low
- Excessive fuel rates and/or insufficient kiln draft

Possible Effects and Dangers:
- Explosion or thermal damage to kiln back-end equipment

Possible Actions:
1. Immediately de-energize electrostatic precipitator.
2. Immediately reduce fuel rate (don't shut off) and increase I.D. fan to obtain:
   a) zero combustibles in exit gas
   b) oxygen between 0.2 and a maximum of 0.5% in exit gas
3. After black smoke has cleared, maintain the low oxygen/zero combustibles for at least 10 min before restoring kiln variables to normal.

Preventive Measures:
- Improve control over flame and firing conditions.
- Make frequent, vigilant observation of fuel rates, gas analysis, flame and kiln-draft conditions during kiln starts and upsets.

EMERGENCY CONDITION 23.7: DISTORTED FLAME SHAPE

Indicators:
- Irregular and unusual flame shape
- Fragmented flame where part of flame impinges on lining near the kiln discharge area.

Possible Effects and Dangers:
- Thermal damage to refractory, kiln shell, and kiln hood
- Red spots on the kiln shell near discharge area
- Thermal damage to nose castings

Possible Actions:
1. Inspect burner pipe for damage.
2. If flame is erratic and severely impinges upon lining near the kiln discharge area: SHUT KILN DOWN IMMEDIATELY.
3. If flame is only slightly distorted: Adjust burner position and primary air flow and schedule burner-pipe repairs for next kiln shutdown.

Preventive Measures:
- Regular inspection and maintenance of the burner pipe during each prolonged kiln shutdown.
- Improved protection (castables, air cooling) for burner pipe.
- Maintain primary air flow for at least 2 h after a kiln has been shut down or pull back burner pipe immediately when kiln is being shut down.
- Investigate possibility of relocating and redesigning burner pipe to eliminate frequent damage to burner.

EMERGENCY CONDITION 23.8:
LOSS OF SECTION OF REFRACTORY LINING

Indicators:
- Loose bricks in clinker bed of burning zone
- Delineated (linear instead of round) red spot on the kiln shell
- Rapid rise in shell temperature

Possible Effects and Dangers:
- Thermal damage and distortion of kiln shell and tire
- Further collapse of large sections of linings

Possible Actions:
1. IMMEDIATELY SHUT DOWN KILN.

Preventive Measures:
- Employ proper refractory installation methods and procedures.
- Make annual checks of kiln alignment and shell ovality.
- Have refractory manufacturer provide uniform shapes and proper expansion allowance for each type of brick.
- Avoid excessive turning when kiln is cold during shutdowns.

EMERGENCY CONDITION 23.9:
COOLER DRIVE OR CLINKER BELT STOPPED

Indicators:
- Cooler overloaded
- Large chunks of coating in cooler
- High undergrate pressures
- High cooler drive amps prior to drive stop
- Clinker transfer chutes plugged

Possible Effects and Dangers:
- Thermal damage to cooler components

Possible Actions:
1. Immediately reduce kiln speed to minimum and attempt to restart clinker belt and/or cooler drive.
2. If drives can not be restarted within 5 min, shut kiln down.

NOTE: *After kiln has been shut down, consider possibility of turning the kiln in less frequent intervals to prevent further overloading of cooler* (kiln still has to be turned periodically nevertheless).

Preventive Measures:
- Know at what amperage the cooler drive is likely to fail and provide alarm for overload.
- Adjust kiln parameters (namely kiln speed) before cooler can become overloaded at the times when heavier feed load is observed in the burning zone.

EMERGENCY CONDITION 23.10:
RED CLINKER AT COOLER DISCHARGE

Indicators:
- High drag-chain amps
- Sudden drop in undergrate pressure (grate out)
- Excessively high undergrate pressure (cooler overloaded)
- Cooler drive amps and clinker bed depth too high
- Cooler loaded with coating and ring fragments
- Stalagmite formation at cooler inlet
- Uneven cross-sectional loading of cooler
- Insufficient air flow into cooler

Possible Effects and Dangers:
- Thermal damage to cooler components
- Thermal damage to clinker transport equipment

Possible Actions:
1. Immediately make a visual check of the cooler to determine reason for red-clinker discharge.
   If cooler grate out, *SHUT KILN DOWN.*
   If cooler overloaded, *REDUCE KILN SPEED TO MINIMUM AND REDUCE COOLER-GRATE DRIVE SPEED TO ALLOW MORE TIME FOR COOLING.*
2. Increase air flow into cooler.
3. Activate water spray at cooler discharge and reroute clinker to prevent damage to conveyor belts.

Preventive Measures:
A) On frequent grate failures:
- Investigate for possible faulty grate-installation methods by maintenance department.
- Investigate quality of grates and bolts used.

B) On frequent one-sided loading of cooler bed:
- Investigate possible cooler-design changes.
- Investigate possibilities for elimination of stalagmite ("snowmen") formation at cooler inlet.

C) On frequent overloading of cooler due to upsets:
- Slow kiln speed down before raw feed enters cooler or cooler can become overloaded (make your corrective moves *before* things get out of control).

EMERGENCY CONDITION 23.11:
RAPID RISE OF TEMPERATURE IN COAL SYSTEM

Possible Effects and Dangers:
- Explosion
- Thermal damage to coal system

Possible Actions:
WARNING: *Do not open any doors in the system that could provide the oxygen for an explosion or more serious fire.*

1. Inject inert gas (CO_2) into coal-mill inlet.
2. Flood coal mill with kiln feed or excessive coal.
3. Warn all personnel to stay clear of system.
4. Stop or reduce air flow to coal mill to minimum.

Preventive Measures for Reoccurence:
- Provide coal-mill inlet with magnetic device to extract metal fragments from coal feeder belt.
- Keep paper, rags, etc. out of coal storage pile.
- Do not feed coal mill with coal that has undergone spontaneous ignition ("smothering") while in storage.
- Keep coal-mill detramp chute clear.
- Provide coal-mill system with automatic fire-extinguishing devices.
- Do not operate coal mill above predetermined safe temperature for any given type of coal.

EMERGENCY CONDITION 23.12:
POWER FAILURE

Possible Effects and Dangers:
- Warpage of kiln shell
- Thermal damage to burner pipe, instrumentation, and equipment at kiln discharge area
- *On Coal-Fired Kilns:* Settlement of ground coal in coal system that could lead to a fire and/or explosion

Possible Actions:
1. *Immediately,* start auxiliary power generator and primary air fan (coal-mill fan on direct-fired kilns).
2. Retract burner pipe and protect T.V. monitor in kiln hood.
3. Start quarter turn on kiln not later than 10 min after the power failure.
4. *If available,* close feed-end damper manually to prevent hot gases from escaping from kiln by natural draft.

EMERGENCY CONDITION 23.13:
CHAIN "FIRE"

Indicators:
- Rapid, sudden rise in intermediate- and exit-gas temperatures
- By visual observation

Possible Effects and Dangers:
- Melt-down and loss of chains
- Damage to kiln shell in chain-system area
- *On wet-process kilns:* Steam explosion
- Thermal damage to kiln back-end equipment

Possible Actions:

WARNING: *Under no circumstances should there be water added at the feed end.*

1. *Immediately,* reduce fuel rate to minimum (BUT DON'T SHUT FUEL OFF COMPLETELY!!!). At the same time, reduce I.D. fan speed to obtain zero combustibles and less than 0.3% oxygen.
2. Increase kiln speed and feed rate to maximum until the back-end temperature is under control.
3. *On wet-process kilns:* Clear all personnel from firing floor.

Preventive Measures:
- Avoid operating the kiln for more than 10 min when there is a feed shortage.
- Establish and enforce maximum permissible operating limits for intermediate and/or exit-gas temperatures.

EMERGENCY CONDITION 23.14:
HEAVY RAIN OR THUNDERSTORM

Possible Effects and Dangers:
On kilns that are exposed to the elements:
- Loss of coating and collapse of refractory lining
- Thermal damage to kiln shell
- Possibility of a power failure

Possible Actions:
If storm occurs shortly after a kiln shutdown:
1. Jack (turn) kiln more frequently or turn continuously on auxiliary drive.
2. Start auxiliary power generator in preparation for a possible power failure.

EMERGENCY CONDITION 23.15:
SUDDEN, HIGH-POSITIVE HOOD PRESSURE

Possible Reasons:
- I.D. fan failure
- Large ring or buildup broken loose inside kiln
- Instrumentation failure of cooler air flow, cooler stack damper, or
- I.D. fan control
- Steam explosion on wet-process kilns

Possible Effects and Dangers:
- All personnel on firing floor is in peril
- Thermal damage to equipment on firing floor and hood
- Danger of back-fire in coal system

Possible Actions:
1. *Immediately,* clear all personnel from firing floor.
2. *Immediately,* reduce fuel rate to minimum and increase I.D. fan speed.
3. Reduce cooler air-flow rates into undergrate compartments.
4. Open cooler excess air damper manually.

24.

Safety and Accident Prevention

The previous chapters have concentrated on kiln control techniques and equipment. To do justice to all aspects of kiln operation, however, a discussion of safety is necessary in a book of this kind. There are many situations in which a worker could be injured because of a lack of machine guards, failure to wear proper protective clothing, or faulty job performance by himself or another person. A kiln operator must familiarize himself with all the potential hazards that might exist in and around the kilns under his control, and set for himself a high standard of safety consciousness. He especially should be alert to point out hazards to other employees and should see to it that no employee works in an unsafe manner on his kilns.

Before entering into a detailed discussion of the hazards around kilns, fundamentals of safety in general should be reviewed so the reader can relate them to the rotary kiln.

24.1   SAFETY

Simply stated, safety measures are introduced into a plant for two reasons: to protect an employee from injury, either physically or financially, while performing his work, and to prevent financial loss to the employer as a result of damage to the equipment or compensation payments which are a part of nearly every industrial accident. Management and employee alike are responsible for making the plant a safe place in which to work, and achieving injury-free work performance day after day. A plant safety program can only be successful when all parties wholeheartedly believe in safety, and when safety becomes a part of the working life of every man in the plant regardless of his position. Evasion of safety responsibilities by the individual, implicitly delegating such responsibility to others, generally referred to as "passing the buck," is bound to result in failure of the program.

If a supervisor appears to be strict and unyielding with respect to safety rules and procedures, his efforts should be appreciated, and not resented. After all, it is the responsibility of the supervisor to see that the employee first endeavors to make himself a safe worker, and only after he has accomplished this and is a good example to others can the employee then try to win others over to the side of safety. That's what safety is all about. It is first of all a state of mind, an idea implemented by a constructive attitude that causes a man to recognize dangerous situations before an accident occurs. It is not something to be lived with under duress because it has been imposed in the form of rules by management. Most importantly, it deserves the support of all employees in the plant.

24.2   ACCIDENTS

Now consider the word "accident." An accident to many workers represents a condition in which someone is injured and property is damaged. Anything less is looked on as a close call, a near miss, or a bit of good luck. To put it in the proper perspective, an accident is an accident even though no one is injured or no damage is done. An accident is any unintentional or unexpected interruption of the orderly progress of the work. Accidents do not happen. An accident is the result of some unsafe act or equipment. We have but little control over the severity of injury once an accident has occurred, but we can control the conditions leading up to the injury. A statistical analysis of thousands of accidents and injuries shows that every accident that resulted in major injury (a lost time accident) was preceded by 29 minor accidents (no lost time and only minor injury) and by 300 accidents that caused no injury at all. So-called near misses and close calls are included in the 300. These statistics warn that, if we have a great number of close calls on the job, sooner or later there will be a serious injury in the plant. Ironically, one usually finds the reasons for an injury accident, but seldom wants to take the time to do something after a close call to prevent these minor accidents from reaching major proportions.

For every accident that is the result of unsafe conditions there are nine that were caused by unsafe acts, including those resulting from failure to recognize unsafe conditions. An employee can easily fall into the habit of overlooking some basic safety procedures and taking unnecessary chances when he develops the attitude that because nothing happened the last time, nothing will happen the next time he does the same thing.

The "Accident Roundtable," published montly by the Portland Cement Association, points out that accidents in the vicinity of kilns have a higher frequency rate than those in other areas of the plant. It is common practice in cement plants to provide general safety rules that apply to all employees throughout the plant. There are, however, certain hazards that are unique to rotary kilns, and it is these dangers that the kiln operator must become aware of. Table 24.1 is a compilation of kiln hazards and possible action that can be taken to eliminate or reduce the dangerous condition.

TABLE 24.1

KILN HAZARDOUS CONDITIONS

Hazard	Action to eliminate or reduce hazard
Backfire and explosion during kiln light-up	• Open either one cooler or burner hood door before lighting fire in kiln. • Secure proper draft in kiln before fire is lighted (very important). • Do not allow unauthorized persons to stand near the burner hood during light-up. • Stay clear of burner hood ports when igniting the fuel. • Avoid excessive fuel flow on initial light-up of the flame. • Start the primary-air fan before opening the fuel valve. • When firing coal, make sure that no coal-dust spills are present on firing floor, around coal feeder, or in the primary-air pipe.

Hazard	Action to eliminate or reduce hazard
Setting any kiln machinery into motion during startup	• Make sure all persons are clear of kiln equipment before each unit is started. • Sound horn (if available) to signal startup. • Inspect all circuit breakers before the startup to make sure that all safety tags and locks have been removed. • Make sure all machine guards are in place before any equipment is started.
Relining the kiln with refractory bricks and material	• Construct a proper bridge across the burner hood from firing floor to kiln nose. • Inspect coating and remove loose overhangs before passing underneath. • Keep all unauthorized personnel out of kiln interior. • Use protective screen when working under loose refractory and coating, if no alternate procedure is possible. • Any employee working inside the kiln should have positive means, such as locking out the kiln drive with his own lock, to assure that the kiln cannot be started while he is inside. • Have proper posture and steady footing when lifting bricks or scaling coating. • Do not work underneath the burner hood bridge while material is being hauled in and out of the kiln. • Do not test run cooler fans when workmen are inside kiln. • Do not run I.D. fan when workmen are at kiln rear or in chain section.
Working near or on dust-collecting equipment	• Wear extra protective clothing to guard against burns from hot dust. • Wash skin thoroughly with clear water after contact with alkaline dust. • Have a second workman as safety man standing by whenever working under or in bins or hoppers containing material. • Do not allow workmen to work inside hoppers without being properly secured on safety lines and belts.

Hazard	Action to eliminate or reduce hazard
	• When working on plugged flue chambers, be constantly on guard against potential dust flushes and cave-in of overhanging material.
Shooting clinker rings with industrial gun	• Do not allow any employees other than the gun crew on the firing floor during ring shooting. • Do not tamper with the ammunition. • Keep all live ammunition locked up and away from the firing floor when not in use. • Permit only experienced and trained persons to operate the kiln gun. • Use ear muffs or plugs when firing gun. • Cotton sutffed in the ear is not adequate. • Clean gun at frequent intervals and do not attempt to fire an apparently defective gun. • If kiln has no chain section, keep all persons away from the kiln back end and rope this area off before shooting.
Clinker, fuel oil, and coal dust spills	• Clean up any spills immediately. • Provide adequate clean-up cans and facilities for easy removal of spills. • Initiate repair action when spills are caused by leaks that can be repaired.
Gas, fuel oil, coal, and steam leaks in fuel system	• Report any gas odor on the firing floor immediately to the foreman. • Provide for periodic inspection of fuel and steam lines and system to detect leaks and other defects as a preventive measure against major breaks in the system.
Burner hood portholes and cooler doors	• Do not allow anyone to look into the burning zone while the kiln is in operation unless approved safety equipment for viewing is used. • Use proper protective clothing when working near open burner hood and cooler doors while the kiln is in operation. • Instruct all persons to stay clear of the portholes whenever the hood pressure is temporarily on the positive side.

Appendix

A: THE INTERNATIONAL SYSTEM OF UNITS (SI)

The following is a guide to familiarize the reader with the units, prefixes, symbols, and formulas used in the International System of Units. "SI" is the common language in which scientific and technical data are presented worldwide.

A.1 Base Units

Quantity	Unit	SI symbol
length	meter	m
mass	kilogram	kg
time	second	s
electric current	ampere	A
thermodynamic temperature	Kelvin	K
amount of substance	mole	mol
luminous intensity	candela	cd

A.2 Supplementary Units

plane angle	radian	rad
solid angle	steradian	sr

A.3 Derived Units

Quantity	Unit	SI symbol	Formula
acceleration	meter per second^2		m/s^2
angular acceleration	radian per second^2		rad/s^2
angular velocity	radian per second		rad/s
area	square meter		m^2
density	kilogram per meter^3		kg/m^3
electric capacitance	farad	F	A·s/V
electric field strength	volt per meter		V/m
electric conductance	siemens	S	A/V
electric inductance	henry	H	V·s/A
electric potential diff.	volt	V	W/A
electric resistance	ohm	Ω	V/A
electromotive force	volt	V	W/A
energy	joule	J	N·m
entropy	joule per kelvin		J/K
force	newton	N	kg·m/s^2
frequency	hertz	Hz	(cycle)/s
illuminance	lux	lx	lm/m^2
luminance	candela per meter^2		cd/m^2
luminous flux	lumen	lm	cd·sr
magnetic field strength	ampere per meter		a/m
magnetic flux	weber	Wb	V·s
magnetic flux density	tesla	T	Wb/m^2
magnetomotive force	ampere	A	
power	watt	W	J/s
pressure	pascal	Pa	N/m^2
quantity of electricity	coulomb	C	A·s
quantity of heat	joule	J	N·m
radiant intensity	watt per steradian		W/sr
specific heat	joule per kg-kelvin		J/kg·K
stress	pascal	Pa	N/m^2
thermal conductivity	watt per m-kelvin		W/m·K
velocity	meter per second		m/s
viscosity dynamic	pascal-second		Pa·s
viscosity, kinematic	m^2 per second		m^2/s
voltage	volt	V	W/A
volume*	cubic meter		m^3
wavenumber	reciprocal meter		(wave)/m
work	joule	J	N·m

* In normal engineering work, where high precision is not required, the use of the liter as a unit to express volume is acceptable.

B: WEIGHTS AND MEASURES

B.1 Weights

1 lb	=	0.4536	kg	1 kg	=	2.2046	lb
1 lb	=	16	oz	1 g	=	0.0352739	oz
1 lb	=	453.59	g	1 g	=	0.0022046	lb
1 lb	=	444820	dynes	1 dynes	=	2.248E–06	lb
1 short ton	=	0.907185	metric ton	1 metric ton	=	1.102311	short ton
1 short ton	=	907.2	kg	1 kg	=	0.0011023	short ton
1 short ton	=	2000	lb	1 metric ton	=	2204.5	lb
1 lb	=	0.0004536	metric ton	1 metric ton	=	1000	kg
1 oz	=	28.3495	g	1 kg	=	35.2739	oz

B.2 Linear Measures

1 in.	=	25.4	mm	1 mm	=	0.03937 in.
1 in.	=	2.54	cm	1 cm	=	0.3937 in.
1 in.	=	0.0254	m	1 m	=	39.370079 in.
1 ft	=	0.3048	m	1 m	=	3.2808399 ft
1 ft	=	30.479	cm	1 cm	=	0.0328095 ft
1 ft	=	300.479	mm	1 mm	=	0.003328 ft
1 ft	=	0.0003048	km	1 km	=	3280.8399 ft
1 stat. mi	=	1.609	km	1 km	=	0.621504 stat. mi
1 stat. mi	=	0.8684	naut. mi	1 km	=	1000 m
1 naut. mile	=	1.1515	stat. mi	1 micron	=	25.4 μm
1 yd	=	3	ft			
1 yd	=	0.9149	m	1 m	=	1.0930156 yd

B.3 Areas

1 in.²	=	6.4516	cm²		
1 ft²	=	144	in.²		
1 ft²	=	0.092903	m²		
1 ft²	=	929.03	cm²		
1 ft²	=	2.296E−05	acres		
1 mi²	=	640	acres		
1 mi²	=	2.59	km²		
1 acre	=	43560	ft²		
1 acre	=	4046.8	m²		
1 yd²	=	0.8361	m²		
1 cm²	=	0.1550003	in.²		
1 cm²	=	0.001076	ft²		
1 m²	=	10.763915	ft²		
1 hectare	=	2.47	acres		
1 are	=	119.6	yd²		
1 km²	=	247.10883	acres		
1 km²	=	0.3861004	mi²		
1 are	=	100	m²		
1 m²	=	0.0002471	acres		
1 m²	=	1.1960292	yd²		

B.4 Volumes

1 in.³	=	16.387	cm³	1 cm³	=	0.061024	in.³
1 ft³	=	28.317	dm³	1 dm³	=	0.0353145	ft³
1 liter	=	1	dm³	1 dm³	=	1	1
1 ft³	=	1728	in.³				
1 yd³	=	0.76455	m³	1 m³	=	1.3079589	yd³
1 fl oz USA	=	0.029574	dm³	1 dm³	=	33.813485	fl oz US
1 pt US	=	0.47318	dm³	1 dm³	=	2.1133607	pt US
1 qt US	=	0.94636	dm³	1 dm³	=	1.0566803	qt US
1 gal US	=	3.78543	dm³	1 dm³	=	0.2641708	gal US
1 bbl oil	=	158.762	dm³	1 dm³	=	0.0062987	bbl. oil
1 yd³	=	27	ft³	1 dm³	=	61.02	in.³

B.5 Specific Weights (Densities)

1 gr/ft³	=	2.288E–06	g/cm³	1 kg/dm³	=	1	g/cm³
1 lb/ft³	=	0.0160185	g/cm³	1 g/cm³	=	436998	gr/ft³
1 lb/yd³	=	0.0005933	g/cm³	1 g/cm³	=	62.427818	lb/ft³
1 lb/in.³	=	2.767997	g/cm³	1 g/cm³	=	1685.4879	lb/yd³
1 lb/ft³	=	16.02	kg/m³	1 g/cm³	=	0.3612721	lb/in.³
1 lb/gal US	=	7.48	lb/ft³	1 kg/m³	=	0.062422	lb/ft³
				1 kg/m³	=	0.01	kg/dm³

B.6 Specific Flow Rates (Velocities)

1 ft³/min	=	0.02832	m³/min		
1 ft³/s	=	448.83	gal US/min		
1 ft/min	=	0.508	cm/s		
1 ft/min	=	0.018288	km/h		
1 ft/min	=	0.3048	m/min		
1 ft/min	=	0.011364	mi/h		
1 gal/min	=	0.22712	m³/h		
1 gal/min	=	0.063088	dm³/s		
1 mi/h	=	88	ft/min		
1 mi/h	=	1.609	km/h		
1 m³/min	=	35.310734	ft³/min		
1 gal US/min	=	0.002228	ft³/s		
1 cm/s	=	1.9685039	ft/min		
1 km/h	=	54.680665	ft/min		
1 m/min	=	3.2808399	ft/min		
1 mi/h	=	87.997184	ft/min		
1 m³/h	=	4.4029588	gal/min		
1 dm³/s	=	15.850875	gal/min		
1 ft/min	=	0.0113636	mi/h		
1 km/h	=	0.621504	mi/h		

B.7 Specific Weights (Gases)

1 lb/SCF dry	=	16.882	kg/m³ dry
1 lb/SCF wet	=	17.078	kg/m³ wet
1 gr/SCF dry	=	0.0024118	kg/m³ dry
1 gr/SCF wet	=	0.0024397	kg/m³ wet
1 kg/m³ dry	=	0.0592347	lb/SCF dry
1 kg/m³ wet	=	0.0585549	lb/SCF wet
1 kg/m³ dry	=	414.62808	gr/SCF dry
1 kg/m³ wet	=	409.88646	gr/SCF wet

SCF = 30 in. Hg @ 60 F

1 N. (normal) = 0 C, 1.01325 bar

B.8 Force

1 poundal	=	0.138255	N		
1 lb-force	=	3.338221	N		
1 kp	=	9.80665	N		

1 N	=	7.2330115	poundal
1 N	=	0.2995608	lb-force
1 N	=	0.1019716	kp

B.9 Pressure
Note: Pa = N/m²

1 psi	=	6.8948	kN/m²		
1 psi	=	68.948	mbar		
1 psi	=	2.3066	ft H₂O		
1 psi	=	0.0703062	kg/cm²		
1 psi	=	703.07	kg/m²		
1 in. Hg	=	0.038638	bar		
1 in. Hg	=	0.03342	atm		
1 in. Hg	=	3863.8	N/m²		
1 in. H₂O	=	0.002539	kg/cm²		
1 lb/ft²	=	0.0004882	kg/cm²		
1 lb/ft²	=	47.876	N/m²		
1 N/m²	=	0.000001	N/mm²		
1 kp/cm²	=	0.0981	N/mm²		
1 kp/cm²	=	98100	N/m²		
1 N	=	0.102	kp		
1 psi	=	6894.76	N/m²		
1 in. H₂O	=	24.899	N/m²		
1 psi	=	0.0068953	N/mm²		

1 kN/m²	=	0.1450368	psi
1 mbar	=	0.0145037	psi
1 ft H₂O	=	0.4335385	psi
1 kg/cm²	=	14.2235	psi
1 kg/m²	=	0.0014223	psi
1 bar	=	25.881257	in. Hg
1 atm	=	29.922202	in. Hg
1 N/m²	=	0.0002588	in. Hg
1 kg/cm²	=	393.85585	in. H₂O
1 lb/ft²	=	2048.3408	lb/ft²
1 N/m²	=	0.0208873	lb/ft²
1 N/mm²	=	1000000	N/m²
1 N/mm²	=	10.204082	kp/cm²
1 N/m²	=	1.019E-05	kp/cm²
1 kp	=	9.8039216	N
1 N/m²	=	0.000145	psi
1 N/m²	=	0.0401623	in. H₂O
1 N/mm²	=	145.02632	psi

1 Btu	=	0.2518892	kcal	
1 Btu	=	1.055056	kJ	
1 Btu/lb F	=	1.00041	kcal/kg C	
1 Btu/lb F	=	4.1886	kJ/kg C	
1 Btu/ft²hF	=	4.8844	kcal/m²hC	
1 Btu/ft²hF	=	20.45	kJ/m²hC	
1 Btu-in./ft²h	=	0.068925	kcal/mh	
1 Btu-in./ft²h	=	0.288578	kJ/mh	
1 Btu/fth	=	0.8271	kcal/mh	
1 Btu/fth	=	3.46294	kJ/mh	
1 Btu-in./ft²hF	=	0.12407	kcal/mhC	
1 Btu-in./ft²hF	=	0.51946	kJ/mhC	
1 Btu/lb	=	0.55579	kcal/kg	
1 Btu/lb	=	2.327	kJ/kg	
1 Btu/ft²	=	2.7136	kcal/m²	
1 Btu/ft²	=	11.361138	kJ/m²	
1 Btu/in.²	=	390.76	kcal/m²	
1 Btu/in.²	=	1636.1	kJ/m²	
1 Btu/short ton	=	0.000278	kcal/kg	
1 Btu/short ton	=	0.0011631	kJ/kg	
1 J	=	0.0002388	kcal	
1 J	=	2.78E–07	kWh	
1 kcal/kg	=	4.184	MJ/ton	
1 kcal/m²h	=	1.163	W/m²	
1 kcal/mhC	=	1.163	W/mK	
1 kcal/m²hC	=	1.163	W/Km²	
1 kcal/kgC	=	4.184	kJ/kgK	

1 kcal	=	3.96999	Btu
1 kJ	=	0.947817	Btu
1 kcal/kg C	=	0.9995902	Btu/lb F
1 kJ/kg C	=	0.2387433	Btu/lb F
1 kcal/m²hC	=	0.2047334	Btu/ft²hF
1 kJ/m²hC	=	0.0488998	Btu/ft²hF
1 kcal/mh	=	14.508524	Btu-in./ft²h
1 kJ/mh	=	3.4652676	Btu-in./ft²h
1 kcal/mh	=	1.2090436	Btu/fth
1 kJ/mh	=	0.288772	Btu/fth
1 kcal/mhC	=	8.0599661	Btu-in./ft²h
1 kJ/mhC	=	1.925076	Btu-in./ft²h
1 kcal/kg	=	1.7992407	Btu/lb
1 kJ/kg	=	0.4297379	Btu/lb
1 kcal/m²	=	0.3685142	Btu/ft²
1 kJ/m²	=	0.0880194	Btu/ft²
1 kcal/m²	=	0.0025591	Btu/in.²
1 kJ/m²	=	0.0006112	Btu/in.²
1 kcal/kg	=	3597.122	Btu/short ton
1 kJ/kg	=	859.7713	Btu/short ton
1 kcal	=	4187	J
1 kWh	=	36000	J
1 MJ/ton	=	0.2390057	kcal/kg
1 W/m²	=	0.8598452	kcal/m²h
1 W/mK	=	0.8598452	kcal/mhC
1 W/Km²	=	0.8598452	kcal/m²hC
1 kJ/kgK	=	0.2390057	kcal/kgC

B.11 Work and Energy
W = J/s

1 ft-lb (weight)	=	1.35582	J	1 J	=	0.737561	ft-lb (weight)
1 ft-lb (force)	=	1.35582	J	1 J	=	0.737561	ft-lb (force)
1 ft-poundal	=	0.0421401	J	1 J	=	23.730366	ft-poundal
1 Btu	=	1.055056	kJ	1 kJ	=	0.947817	Btu
1 Btu	=	0.293071	Wh	1 Wh	=	3.4121425	Btu
1 hp-h	=	2.68452	MJ	1 MJ	=	0.3725061	hp-h
1 hp-h	=	0.7457	kWh	1 kWh	=	1.3410219	hp-h
1 ft-lb (wt)/s	=	1.35582	W	1 W	=	0.737561	ft-lb (wt)/s
1 Btu	=	1.055056	kW	1 kW	=	0.947817	Btu
1 Btu/h	=	0.29307	kW	1 kW	=	3.4121541	Btu/h
1 hp	=	0.7457	kW	1 kW	=	1.3410219	hp
1 PS	=	735.5	W	1 W	=	0.0013596	PS
1 kWh	=	860	kcal	1 kcal	=	0.0011628	kWh
1 W	=	0.86	kcal/h	1 kcal/h	=	1.1627907	W
1 W	=	0.001	kW	1 kW	=	1000	W

B.12 Engineering Units

tera	T	trillion	$1 \times E$ 12	
giga	G	billion	$1 \times E$ 9	1000000000
		hundred million	$1 \times E$ 8	100000000
		ten million	$1 \times E$ 7	10000000
mega	M	million	$1 \times E$ 6	1000000
		hundred thousand	$1 \times E$ 5	100000
		ten thousand	$1 \times E$ 4	10000
kilo	k	thousand	$1 \times E$ 3	1000
hecto	h	hundred	$1 \times E$ 2	100
deca	da	ten	$1 \times E$ 1	10
			$1 \times$	1
deci	d	tenth	$1 \times E$ −1	0.1
centi	c	hundredth	$1 \times E$ −2	0.01
milli	m	thousandths	$1 \times E$ −3	0.001
		ten thousandths	$1 \times E$ −4	0.0001
		hundred thousandths	$1 \times E$ −5	0.00001
micro	μ	millionth	$1 \times E$ −6	0.000001
nano	n	billionth	$1 \times E$ −9	0.000000001
pico	p	trillionth	$1 \times E$−12	

C: TEMPERATURE CONVERSIONS

C		F	C		F	C		F
−34	−30	−22	−12	11	52	11	52	126
−34	−29	−20	−11	12	54	12	53	127
−33	−28	−18	−11	13	55	12	54	129
−33	−27	−17	−10	14	57	13	55	131
−32	−26	−15	−9	15	59	13	56	133
−32	−25	−13	−9	16	61	14	57	135
−31	−24	−11	−8	17	63	14	58	136
−31	−23	−9	−8	18	64	15	59	138
−30	−22	−8	−7	19	66	16	60	140
−29	−21	−6	−7	20	68	16	61	142
−29	−20	−4	−6	21	70	17	62	144
−28	−19	−2	−6	22	72	17	63	145
−28	−18	0	−5	23	73	18	64	147
−27	−17	1	−4	24	75	18	65	149
−27	−16	3	−4	25	77	19	66	151
−26	−15	5	−3	26	79	19	67	153
−26	−14	7	−3	27	81	20	68	154
−25	−13	9	−2	28	82	21	69	156
−24	−12	10	−2	29	84	21	70	158
−24	−11	12	−1	30	86	22	71	160
−23	−10	14	−1	31	88	22	72	162
−23	−9	16	0	32	90	23	73	163
−22	−8	18	1	33	91	23	74	165
−22	−7	19	1	34	93	24	75	167
−21	−6	21	2	35	95	24	76	169
−21	−5	23	2	36	97	25	77	171
−20	−4	25	3	37	99	26	78	172
−19	−3	27	3	38	100	26	79	174
−19	−2	28	4	39	102	27	80	176
−18	−1	30	4	40	104	27	81	178
−18	0	32	5	41	106	28	82	180
−17	1	34	6	42	108	28	83	181
−17	2	36	6	43	109	29	84	183
−16	3	37	7	44	111	29	85	185
−16	4	39	7	45	113	30	86	187
−15	5	41	8	46	115	31	87	189
−14	6	43	8	47	117	31	88	190
−14	7	45	9	48	118	32	89	192
−13	8	46	9	49	120	32	90	194
−13	9	48	10	50	122	33	91	196
−12	10	50	11	51	124	33	92	198

C		F	C		F	C		F
34	93	199	57	135	275	81	177	351
34	94	201	58	136	277	81	178	352
35	95	203	58	137	279	82	179	354
36	96	205	59	138	280	82	180	356
36	97	207	59	139	282	83	181	358
37	98	208	60	140	284	83	182	360
37	99	210	61	141	286	84	183	361
38	100	212	61	142	288	84	184	363
38	101	214	62	143	289	85	185	365
39	102	216	62	144	291	86	186	367
39	103	217	63	145	293	86	187	369
40	104	219	63	146	295	87	188	370
41	105	221	64	147	297	87	189	372
41	106	223	64	148	298	88	190	374
42	107	225	65	149	300	88	191	376
42	108	226	66	150	302	89	192	378
43	109	228	66	151	304	89	193	379
43	110	230	67	152	306	90	194	381
44	111	232	67	153	307	91	195	383
44	112	234	68	154	309	91	196	385
45	113	235	68	155	311	92	197	387
46	114	237	69	156	313	92	198	388
46	115	239	69	157	315	93	199	390
47	116	241	70	158	316	93	200	392
47	117	243	71	159	318	94	201	394
48	118	244	71	160	320	94	202	396
48	119	246	72	161	322	95	203	397
49	120	248	72	162	324	96	204	399
49	121	250	73	163	325	96	205	401
50	122	252	73	164	327	97	206	403
51	123	253	74	165	329	97	207	405
51	124	255	74	166	331	98	208	406
52	125	257	75	167	333	98	209	408
52	126	259	76	168	334	99	210	410
53	127	261	76	169	336	99	211	412
53	128	262	77	170	338	100	212	414
54	129	264	77	171	340	101	213	415
54	130	266	78	172	342	101	214	417
55	131	268	78	173	343	102	215	419
56	132	270	79	174	345	102	216	421
56	133	271	79	175	347	103	217	423
57	134	273	80	176	349	103	218	424

C		F	C		F	C		F
104	219	426	152	305	581	268	515	959
104	220	428	154	310	590	271	520	968
105	221	430	157	315	599	274	525	977
106	222	432	160	320	608	277	530	986
106	223	433	163	325	617	279	535	995
107	224	435	166	330	626	282	540	1004
107	225	437	168	335	635	285	545	1013
108	226	439	171	340	644	288	550	1022
108	227	441	174	345	653	291	555	1031
109	228	442	177	350	662	293	560	1040
109	229	444	179	355	761	296	565	1049
110	230	446	182	360	680	299	570	1058
111	231	448	185	365	689	302	575	1067
111	232	450	188	370	698	304	580	1076
112	233	451	191	375	707	307	585	1085
112	234	453	193	380	716	310	590	1094
113	235	455	196	385	725	313	595	1103
113	236	457	199	390	734	316	600	1112
114	237	459	202	395	743	318	605	1121
114	238	460	204	400	752	321	610	1130
115	239	462	207	405	761	324	615	1139
116	240	464	210	410	770	327	620	1148
116	241	466	213	415	779	329	625	1157
117	242	468	216	420	788	332	630	1166
117	243	469	218	425	797	335	635	1175
118	244	471	221	430	806	338	640	1184
118	245	473	224	435	815	341	645	1193
119	246	475	227	440	824	343	650	1202
119	247	477	229	445	833	349	660	1220
120	248	478	232	450	842	354	670	1238
121	249	480	235	455	851	360	680	1256
121	250	482	238	460	860	366	690	1274
124	255	491	241	465	869	371	700	1292
127	260	500	243	470	878	377	710	1310
129	265	509	246	475	887	382	720	1328
132	270	518	249	480	896	388	730	1346
135	275	527	252	485	905	393	740	1364
138	280	536	254	490	914	399	750	1382
141	285	545	257	495	923	404	760	1400
143	290	554	260	500	932	410	770	1418
146	295	563	263	505	941	416	780	1436
149	300	572	266	510	950	421	790	1454

C		F	C		F	C		F
427	800	1472	660	1220	2228	893	1640	2984
432	810	1490	666	1230	2246	899	1650	3002
438	820	1508	671	1240	2264	904	1660	3020
443	830	1526	677	1250	2282	910	1670	3038
449	840	1544	682	1260	2300	916	1680	3056
454	850	1562	688	1270	2318	921	1690	3074
460	860	1580	693	1280	2336	927	1700	3092
466	870	1598	699	1290	2354	932	1710	3110
471	880	1616	704	1300	2372	938	1720	3128
477	890	1634	710	1310	2390	943	1730	3146
482	900	1652	716	1320	2408	949	1740	3164
488	910	1670	721	1330	2426	954	1750	3182
493	920	1688	727	1340	2444	960	1760	3200
499	930	1706	732	1350	2462	966	1770	3218
504	940	1724	738	1360	2480	971	1780	3236
510	950	1742	743	1370	2498	977	1790	3254
516	960	1760	749	1380	2516	982	1800	3272
521	970	1778	754	1390	2534	988	1810	3290
527	980	1796	760	1400	2552	993	1820	3308
532	990	1814	766	1410	2570	999	1830	3326
538	1000	1832	771	1420	2588	1004	1840	3344
543	1010	1850	777	1430	2606	1010	1850	3362
549	1020	1868	782	1440	2624	1021	1870	3398
554	1030	1886	788	1450	2642	1032	1890	3434
560	1040	1904	793	1460	2660	1043	1910	3470
566	1050	1922	799	1470	2678	1054	1930	3506
571	1060	1940	804	1480	2696	1066	1950	3542
577	1070	1958	810	1490	2714	1077	1970	3578
582	1080	1976	816	1500	2732	1088	1990	3614
588	1090	1994	821	1510	2750	1099	2010	3650
593	1100	2012	827	1520	2768	1110	2030	3686
599	1110	2030	832	1530	2786	1121	2050	3722
604	1120	2048	838	1540	2804	1132	2070	3758
610	1130	2066	843	1550	2822	1143	2090	3794
616	1140	2084	849	1560	2840	1154	2110	3830
621	1150	2102	854	1570	2858	1166	2130	3866
627	1160	2120	860	1580	2876	1177	2150	3902
632	1170	2138	866	1590	2894	1188	2170	3938
638	1180	2156	871	1600	2912	1199	2190	3974
643	1190	2174	877	1610	2930	1210	2210	4010
649	1200	2192	882	1620	2948	1221	2230	4046
654	1210	2210	888	1630	2966	1232	2250	4082

C		F	C		F	C		F
1243	2270	4118	1388	2530	4586	1532	2790	5054
1254	2290	4154	1399	2550	4622	1543	2810	5090
1266	2310	4190	1410	2570	4658	1554	2830	5126
1277	2330	4226	1421	2590	4694	1566	2850	5162
1288	2350	4262	1432	2610	4730	1577	2870	5198
1299	2370	4298	1443	2630	4766	1588	2890	5234
1310	2390	4334	1454	2650	4802	1599	2910	5270
1321	2410	4370	1466	2670	4838	1610	2930	5306
1332	2430	4406	1477	2690	4874	1621	2950	5342
1343	2450	4442	1488	2710	4910	1632	2970	5378
1354	2470	4478	1499	2730	4946	1643	2990	5414
1366	2490	4514	1510	2750	4982	1654	3010	5450
1377	2510	4550	1521	2770	5018	1666	3030	5486

D: KILN OPERATOR'S QUIZ

This quiz is designed to give the reader an opportunity to evaluate his knowledge of kiln operation. The quiz consists of 50 true-or-false questions.

D.1 FUNDAMENTALS (Score 5 points for each correct answer)

		T	F

1. When air is heated, its weight per unit volume decreases.   ____ ____
2. One ton of kiln feed will make one ton of clinker.   ____ ____
3. High free-lime contents in the clinker always indicate that the clinker has been underburned.   ____ ____
4. C_3S (tricalcium silicate) is an ingredient in the kiln feed.   ____ ____
5. CO_2 (carbon dioxide) in the kiln exit gases originates partly from the combustion of the fuel and partly from calcination of feed.   ____ ____
6. An increase in fuel rate will always generate more heat in the burning zone.   ____ ____
7. Opening a butterfly damper from 80% to 100% open will most likely produce no increase in flow.   ____ ____

8. Iron and alumina are fluxing agents in the kiln feed mix and tend to make the clinker easier to burn. ___ ___

9. On coal-fired kilns, the chemical composition of the clinker will be the same as the composition of the kiln feed. ___ ___

10. Large changes in the silica ratio will indicate possible changes in the burnability of the clinker. ___ ___

11. A lime saturation factor of 1.03 and a free lime of 1.4% in the clinker means the clinker has been underburned. ___ ___

12. The predominant component in the kiln feed is calcium carbonate. ___ ___

13. Calcium oxide (CaO) is the product when limestone has been calcined. ___ ___

14. All metal components of the kiln, when heated, will contract. ___ ___

15. Higher undergrate pressures in the cooler compartments indicate a higher air-flow rate. ___ ___

16. If the fan speed and damper setting remain the same, the amperes on an air fan will increase when the air temperature decreases. ___ ___

17. The oxygen content in the kiln exit gases is a sole function of the amount of air passing through the kiln. ___ ___

18. Red color on the kiln shell means there is no refractory lining left. ___ ___

19. Higher bed depth in the cooler produces better cooling of the clinker. ___ ___

20. A higher kiln speed shortens the residence time of the feed in the kiln. ___ ___

D.2 OPERATIONAL
(Note: Assume all other kiln variables remain constant.)
(Score 2 points for each correct answer)

                                                             T   F

21. An increase in fuel rate will cause a decrease in the kiln exit-gas oxygen content. ___ ___

22. Slight percentages of combustibles in the kiln exit gas can be ignored. ___ ___

23. Ideally, a kiln operates with approximately 5% excess   ——   ——
    air.

24. In normal operation, the kiln exit gas should show   ——   ——
    approximately 3% oxygen.

25. An increase in I.D. fan speed will increase the   ——   ——
    burning-zone temperature.

26. The burning-zone temperature can only be increased   ——   ——
    by increasing the fuel rate.

27. A decrease in I.D. fan speed will result in a decrease   ——   ——
    in burning-zone temperature.

28. When a ring is building up in the kiln, the feed-end   ——   ——
    draft will increase.

29. The amount of air entering the kiln proper is   ——   ——
    governed by the amount of air that is forced into the
    cooler.

30. Opening the cooler stack damper means less air will   ——   ——
    enter the kiln proper.

31. A positive hood pressure is solely caused by too   ——   ——
    much air being forced into the cooler.

32. A drop in kiln speed will cause the feed-end   ——   ——
    temperature to rise.

33. A reduction in feed rate will cause the feed-end   ——   ——
    temperature to fall.

34. One ton of kiln dust returned will produce the same   ——   ——
    amount of clinker as one ton of fresh kiln feed.

35. It is generally easier to form coating on coal- than gas- ——   ——
    fired kilns.

36. Changing from natural gas to coal firing will not   ——   ——
    alter the chemical composition and burnability of the
    resultant clinker.

37. Increasing the cooler-grate drive speed will bring   ——   ——
    about a thinner clinker bed.

38. An increase in the cooler exhaust-fan speed (or   ——   ——
    opening the damper at the cooler exhaust stack) will
    cause a decrease in the hood pressure.

39. A kiln will always operate better at low production   ——   ——
    rates than at high rates.

40. An increase in the feed ratio will result in higher   ——   ——
    production rates.

41. Frequent kiln stops and upsets tend to shorten the ____ ____
    refractory life.
42. A rotary kiln is never perfectly round. ____ ____
43. At the beginning of a shift, the operator should ____ ____
    always change the controller settings to confrom to
    the settings as they were 24 h earlier when the kiln
    operated in stable fashion.
44. An increase in feed rate will result in a lower feed-end ____ ____
    temperature.
45. Clinker, when burned at a higher temperature, will ____ ____
    show a higher density.
46. When the oxygen in the exit gases shows 0.4% and ____ ____
    the kiln is cooling down, the operator should increase
    the fuel rate.
47. When starting a kiln, the primary-air and I.D. fans ____ ____
    must always be started first before fuel is given to the
    kiln.
48. One should always try first to secure stable operations ____ ____
    before an attempt is made to push the kiln to
    maximum production.
49. Too much air flow into the cooler can impair clinker ____ ____
    advancement through the cooler.
50. The only way one can increase the production rate of ____ ____
    the kiln is to increase the kiln speed.

ANSWERS TO QUIZ

1.	T	11.	F	21.	T	31.	F	41.	T
2.	F	12.	T	22.	F	32.	T	42.	T
3.	F	13.	T	23.	T	33.	F	43.	F
4.	F	14.	F	24.	F	34.	F	44.	T
5.	T	15.	F	25.	F	35.	T	45.	T
6.	F	16.	T	26.	F	36.	F	46.	F
7.	T	17.	F	27.	F	37.	T	47.	T
8.	T	18.	F	28.	T	38.	T	48.	T
9.	F	19.	F	29.	F	39.	F	49.	T
10.	T	20.	T	30.	F	40.	T	50.	F

Index